JIMD Reports
Volume 1

SSIEM

JIMD Reports
Case and Research Reports, 2011/1

Springer

SSIEM

Editor
Society for the Study of Inborn
Errors of Metabolism
c/o ACB
Tooley St 130-132
SE1 2TU London
United Kingdom

ISSN 2192-8304
ISBN 978-3-642-17707-1 e-ISBN 978-3-642-17708-8
DOI 10.1007/978-3-642-17708-8
Springer Heidelberg Dordrecht London New York

Library of Congress Control Number: 2011934753

Printed on acid-free paper

Springer is part of Springer Science+Business Media (www.springer.com)

Contents

JIMD Reports
DOI: 10.1007/8904_2011_8

Received: 7 May 2010; Revised: 16 December 2010;
Accepted: 22 December 2010; Published online: 22 June
Communicated by: Peter Burgard

Psychosocial Aspects of Predictive Genetic Testing for Acute Intermittent Porphyria in Norwegian Minors

Janice Andersen, Sverre Sandberg, Maalfrid Raaheim, and Eva Gjengedal

Abstract *Objective*: The Norwegian Porphyria Centre routinely offers genetic counselling and predictive genetic testing in families diagnosed with porphyria. The aim of this study was to investigate the subjective experiences of adolescents and young adults who were genetically tested for acute intermittent porphyria (AIP) as minors. What were the psychosocial consequences and how were these handled?

Methods: Qualitative interviews of ten Norwegians aged 16–21 years were performed and analysed based on interpretive description. All participants were initially predictively tested for AIP as minors, but three had subsequently developed manifest disease.

Results: The participants considered early diagnosis and lifestyle moderation advantageous, but finding motivation for precaution was difficult. AIP inflicted few psychosocial challenges and was a small part of the participants' identity, but risk of manifest disease was, nevertheless, a cause for concern for two participants with latent AIP. The participants were content with their present level of knowledge and they felt capable of obtaining relevant information when needed. AIP was experienced as a vague condition, and participants and their relatives attributed a variety of symptoms to the disease.

Conclusion and implications: Being genetically tested as a minor was experienced as useful and entailed relatively few adverse psychosocial consequences, although there was a potential for concern. Appropriate and individually tailored genetic counselling and written consent is subsequently advised. What constitutes a suitable age for testing will differ from individual to individual, but these results suggest that parents in collaboration with their children may be suited to decide what age is appropriate.

Keywords Acute intermittent porphyria · Genetic counselling · Minors · Predictive genetic testing · Psychosocial · Qualitative methods

Introduction

Acute Intermittent Porphyria

Porphyrias are rare and mainly inherited diseases caused by reduced activity of different enzymes in the heme synthetic pathway. Each enzyme defect causes a specific disease. These enzyme defects may result in elevated levels of porphyrins or porphyrin precursors leading to a wide range of symptoms. Acute intermittent porphyria (AIP) is one of the most common forms of porphyria. Prevalence of the symptomatic disease is estimated to 1–2 per 100,000 (Herrick and McColl 2004). Attacks are characterised by abdominal pain, vomiting, headache, muscle aches, muscle weakness, pareses, hypertension, tachycardia, mental symptoms and respiratory paralysis (Sassa 2006). AIP has been referred to as "the little imitator" because sometimes non-specific symptoms are easily mistaken for other subjective health complaints (Crimlisk 1997). A range of medications can trigger symptoms, in addition to hormones, alcohol, physical and psychological stress, hunger and fasting (Normann and Puy 2002). Inheritance is autosomal dominant, with reduced

J. Andersen (✉) and S. Sandberg
The Norwegian Porphyria Centre (NAPOS), Laboratory of Clinical Biochemistry, Haukeland University Hospital, Bergen 5021, Norway
and
Department of Public Health and Primary Health Care, University of Bergen, Bergen, Norway
e-mail: janice.andersen@helse-bergen.no

M. Raaheim
Department of Public Health and Primary Health Care, University of Bergen, Bergen, Norway

E. Gjengedal
Department of Public Health and Primary Health Care, University of Bergen, Bergen, Norway
and
Faculty of Health and Social Care, Molde University College, Molde, Norway

penetrance (Badminton and Elder 2005). Disease severity varies from minor symptoms of abdominal pain to serious or frequent attacks requiring hospitalisation. Paralysis and respiratory failure are rare but potentially lethal complications, highlighting the need for early diagnosis (Herrick and McColl 2004). Symptoms very rarely develop before puberty, but there exists a few documented cases of AIP in children (Sandberg et al. 2001; Elder 1997; Hultin et al. 2003).

Genetic Counselling and Testing

The Norwegian Porphyria Centre (NAPOS) offers both diagnostic and predictive genetic testing for AIP. Norwegian legislation states that predictive genetic testing requires extensive genetic counselling, and written informed consent is mandatory (Norwegian Biotechnology Act of 5th December 2003, No. 100 relating to the application of biotechnology in human medicine, etc.). As a rule, predictive genetic testing of minors is not permitted unless such testing can facilitate treatment or prevention of the disease. In the case of AIP, one considers that early detection of mutation carriers will improve prognosis through lifestyle changes and avoidance of symptom-triggering medications. Based on this, NAPOS offers genetic counselling and predictive testing for minors. Mutation carriers without symptoms are referred to as latent AIP, and persons who have experienced one or more AIP attacks are diagnosed as manifest AIP. If a person is diagnosed with latent or manifest AIP, NAPOS will issue a personal ID card that can be used when the person is in contact with health services, and thus contribute to both avoidance of porphyrinogenic medication and the implementation of correct medical treatment.

Previous Research

There are few papers on psychosocial aspects of AIP and hardly any focus on young persons with the diagnosis. A qualitative study from Sweden, in which five women were interviewed on their experiences with serious AIP attacks, concluded that when these women had their attacks they lived in the "deepest darkness", with indescribable pain, both physically and mentally (Wikberg et al. 2000). A British questionnaire study of 116 patients (Millward et al. 2001) found lower quality of life in patients with AIP, compared to patients with other types of porphyria. This resulted in lifestyle consequences for a significant number of the participants and especially for persons with manifest AIP. A follow-up study confirmed that depression and especially anxiety were more common in persons with AIP than in the normal population (Millward et al. 2005). These reports indicate that AIP might have serious consequences for those involved and that there is a need for further investigations of the psychosocial aspects of this disease. Borry et al. (2006) analysed 27 papers from 1991 to 2005 that addressed the issue of guidelines for predictive testing of minors. The main reason for recommending such testing was the direct benefit the patient could achieve through medical treatment or prevention of disease. Unless there were pressing medical reasons for testing, all 27 papers proposed that one should postpone testing until the child could consent. There are few reports concerning predictive testing in minors for adult-onset conditions (Borry et al. 2008; Wade et al. 2010), and it is important to investigate whether young persons benefit from the information such testing yields, and whether risk information causes psychosocial harm to young persons genetically tested for AIP.

Objective

The aim of this study was to investigate the subjective experiences of young persons diagnosed with active or latent AIP as minors. What were the psychosocial consequences, and how were these handled?

Methods

Interpretive descriptive method was chosen as it is a qualitative approach suitable for smaller scale studies seeking clinical insight into subjective experiences (Thorne et al. 1997, 2004). Both the Norwegian Regional Ethics Committee and the Norwegian Social Science Data Services approved this study.

Recruitment and Sample

NAPOS recruited persons eligible for participation. They were contacted through mail, and asked to return written consent to participation, in October 2008 and February 2009. If the participant was younger than 18 years, written parental consent was also required. To meet the inclusion criteria, participants had to be registered at NAPOS, have been genetically tested as minors ($<$ 18 years) and identified as AIP-mutation carriers. It was deemed purposeful that participants were old enough to reflect on the subject in question, but at the same time there should be closeness in time to the genetic testing. Based on this, a theoretical sample consisting of persons not younger than 16 years and

Table 1 Presentation of participants

Participant	Age at PGT[a]	Age at interview	Manifest or latent AIP	Disease manifestation in patients and relatives
1. Alex	10–12	21	lAIP	First-degree relative with unspecific symptoms
2. Morgan	11	17	lAIP	First-degree relative with one AIP attack. Second-degree relative with chronic and unspecific complaints
3. Billie	14	19	mAIP	Unspecific symptoms, more or less chronic complaints. Admitted to hospital twice. First-degree relative with unspecific and frequent complaints in earlier years
4. Drew	5–7	17	mAIP	Three AIP attacks requiring hospitalisation
5. Taylor	17	21	lAIP	No symptomatic relatives in immediate family
6. Chris	12	17	lAIP	First-degree relative with unspecific and frequent complaints
7. Morgan	15	16	lAIP	First-degree relative with several AIP attacks requiring hospitalisation
8. Bobbie	11	19	mAIP	One serious AIP attack. First-degree relative with a few attacks treated at home
9. Cameron	17	21	lAIP	First-degree relative with one attack
10. Francis	15	17	lAIP	First-degree relative with both unspecific and frequent symptoms and attacks requiring medical treatment

Unspecific symptoms – Symptoms perceived by the *participants*: fatigue, muscle aches, headaches, dizziness, etc that were not in combination with otherwise unexplained episodes of stomach/back pain with a duration of >12 h

[a]*PGT* predictive genetic test

not older than 21 years was chosen. Persons with other serious medical conditions were excluded. Twenty-eight persons met the criteria and all were asked to participate. Twelve persons were positive to participation, but as one was difficult to get in touch with, and another was abroad, only ten persons were able to attend. All participants were initially diagnosed with latent AIP, but at the time of the interviews three had experienced one or more attacks and were diagnosed with manifest AIP. A short introduction of the participants is provided in Table 1 to contextualise statements. Analysis did not reveal gender-specific reactions, and participants are given gender-neutral pseudonyms. Relatives are invariably presented as males.

Data Collection

Individual qualitative interviews of six Norwegian females and four males were performed in the period of December 2008 to April 2009 in participants' homes or at a hospital in their proximity. The first author, a registered nurse and genetic counsellor with previous experience from qualitative interviews conducted all interviews. One participant had received genetic counselling from the interviewer in connection to prior genetic testing, without this seeming to influence given answers. Each interview lasted between 40 and 65 minutes and was based on a semi-structured interview guide containing eight topics with a series of open-ended sub-questions. Topics are listed in Table 2. All participants were asked the same questions, but not necessarily in the same order and depth. There was room for the participants to define new topics of interest. Interviews were audiotaped and transcribed verbatim consecutively.

Table 2 Main topics off interview guide

1. Disease presentation in the family
2. Participants own disease
3. Prevention of disease
4. Psychosocial consequences
5. Knowledge of porphyria
6. Being diagnosed as a minor
7. Inheritable disease
8. Genetic counselling

Data Analysis and Interpretative Descriptive Method

Data analysis was performed by the first author, and started as initial readings of the text to get a sense of the whole. For practical purposes, formal analysis commenced after completion of eight interviews, although guidelines for interpretive descriptive approach suggest simultaneous data collection and analysis (Thorne et al. 1997, 2004). There were, nonetheless, certain amounts of informal analysis in the form of reflections in the wake of each interview. A second reading identified nine categories that were used as basis for further coding (Morse 2008). For this coding, the data material was transferred into QSR N6 software (QSR. NUD*IST. Qualitative Research & Solutions Pty Ltd., 1997) for easier data management. Each category was then analysed internally, and summaries were written to more clearly present the participants' meanings and experiences. Based on critical reflection and discussions between the first and the last author, four main themes emerged. In addition, the research questions and what might be clinically relevant and interesting knowledge were of crucial importance. To validate and ensure that participant's initial meaning was not lost, the total text was re-read and interviews were listened to again.

Results

Results are presented as four main themes. Quotes are used to illustrate and validate the participant's meanings.

Living with a Small Intruder

When participants spoke of their genetic status they all perceived knowledge of this as beneficial and useful. In their view, early genetic testing contributed to them stay healthy, and was not experienced as a cause for unnecessary worry. Although risk information was considered advantageous, nine participants were quite clear on the fact that their genetic status did not entail any change of plans in regards to education, occupation and social or physical activities. One participant with frequent and unspecific complaints of AIP contrasted this. The disease made Billie unable to take part in social activities and attend school the way peers did. Because of this, some friends and teachers gave the impression they believed Billie to be lazy.

Participants seemed ambivalent towards lifestyle modifications for prevention of manifest disease, but they would check medication before use. Some claimed to take it easy and to avoid stress, or stated that they exercised and were careful to eat regularly and sleep sufficiently. Nevertheless, they admitted they probably would lead the same life regardless of their genetic status. Cameron had latent AIP. When asked about strategies to avoid manifest AIP, the answer was:

> I used to smoke but I recently cut down on that because I've been thinking smoking is not a smart thing to do when one has porphyria. That really is the only thing I've... But then again, I don't want cancer either, you know? So there are several reasons for me to cut down.

One participant stated that AIP was not a disease, but only a "small intruder" in life. Alex had chosen a career that might be unsuitable for a person with AIP. Several members of the family had asked about the amount of physical or psychological stress the chosen career entailed. Had the risk of triggering an acute attack been considered? The answer to this was:

> It may come as it may, the dreams are bigger than the worries.

Participants did not change their lifestyles *because* of manifest disease or *in fear* of developing manifest disease. Because AIP was something several had never experienced, it was not a big motivator for precaution. Some said they probably ought to do more, but that this was not easy, Taylor said:

> I am just not able to take precautions against something I've never really felt on my body. I might regret not doing it, but up until now I've lived more against the advise I have got, and it has worked out well so far.

Worrying About and Planning for the Future

It was evident that for two participants risk of disease was not unproblematic. They both had latent AIP, and expressed concerns for the future, and worried about the possibility of manifest AIP. They had relatives with chronic complaints attributed to AIP, in addition to other medical conditions. Morgan said:

> I do go around thinking about it all the time, and then I watch my relative and listen to what he has experienced, and I look at my other relative and I'm not supposed to stress (...). I think about it just about every single day

These two participants were quite certain that manifest AIP would invariably lead to disability but this did not affect their choices for the future. Morgans' latent porphyria had not affected choice of education, but when asked about the potential for manifest disease the answer was:

> It is a scary thought, in regards to it being permanent, and what I will actually be able to do afterwards. Some say I will not be allowed to stress as much if I have the disease. Will I be able to work with children like I want to, or? I'm very sceptical of that happening... I believe that at the very least I will not be able to work as much as I would like. And it would be a great shame if I were not able to work with children.

Also, one participant with more frequent AIP complaints had never heard of anybody with manifest disease who was able to work full time. This did not affect the choice of future career, but led to an education where student life was flexible and involved Internet-based lectures.

There were a few reports of discrimination based on genetic status. One participant had been refused acceptance to an education because of latent AIP. Two participants were declared medically unfit for military service because of latent AIP. One had freely provided this information, the other had not. To be declared unfit for service was not experienced as negative, on the contrary one of them said: "In this regard I owe the porphyria a big thank you".

Seeking Information as Needed

Participants believed that information about porphyria was important to their health, but at their current state in life they did not have a pressing need for more information. Cameron said:

> I probably don't know enough, but what I know is sufficient. That is, I know as much as I need to know for now. But if I were to experience an attack I would probably want to know more about it. But right now I don't really feel I need to know any more

Participants generally knew where to get additional information. Most felt that both NAPOS and the two Norwegian patient organisations could be contacted, and also the Internet was frequently used. In addition, the ID card provided by NAPOS was deemed useful. Seven participants claimed to always carry the ID card with them, and some participants reported the card to be their main source of information.

Most participants reported their own age for genetic testing as appropriate, but some believed they had been too young. They had been tested at different ages, but reflected on the fact that they had not been able to understand or care about the implications. Drew said:

> I was very young at the time. But I don't think there was anything wrong with that, really. It was OK. I just didn't understand very much, and it would have been different if I had been told now. I would probably be more curious about it now

AIP a Vague Condition

Most participants perceived AIP as a rather vague disease that they did not have first-hand knowledge of. Contributing to this opinion was individual and ambiguous symptoms and the fact that an attack could be triggered, and manifests itself in varying degrees. Participants also commented that medical authorities could not always offer precise answers regarding AIP. These factors combined implied that there would always be a degree of uncertainty with regard to AIP. When asked whether it was difficult to know what was a symptom of porphyria, Chris answered:

> Yes, I do feel that, you might say. That it can be mixed up in so many things, that you can not clearly separate what is what, especially when I do not have it myself so to speak. If I suddenly get it one day, it is not certain I will recognise it right away. I might just think it is something else

When the participants were asked to describe porphyria in their family, it became evident that a variety of symptoms were attributed to AIP. If a family member said his or her health complaints were caused by porphyria then the participants accepted this as the cause. Participants knew that medications could trigger an acute attack but the mechanisms involved seemed to be vaguely understood. Some explained present health problems in relatives were caused by medication taken years back. Even though participants defined themselves as having *latent* AIP or not having porphyria *yet*, they would attribute symptoms such as feeling tired and "out of it" to AIP. Most participants did not reflect upon this seeming contradiction, Taylor discussed this as follows:

> Because I think, or, I have spent a lot of time becoming certain of what I can attribute to it. And there have been things that would be natural to assume was caused by porphyria but which I think, no. I cannot blame everything on that. I am often tired and so on, but I think a lot of people are, and if I start blaming porphyria for being tired or exhausted then that will just provide me with an excuse, instead of doing something about it.

Discussion

In general, the participant's lives were not very influenced by porphyria, and AIP was only a small part of their identity. One participant, with more chronic complaints of manifest AIP, contrasted this and two young participants with latent AIP worried about their genetic predisposition and were certain that active disease would entail serious consequences. Apart from the participant with chronic complaints, few lifestyle modifications were reported, and although participants believed they would benefit from taking precautions, these were not enough to motivate them into taking active steps to prevent AIP attacks. This ambivalence towards taking precautions is perhaps not surprising, and poses a general problem in health promotion (Shiloh and Ilan 2005; Pronk et al. 2004). That adolescents engage in more risk-taking than adults have previously led to the hypothesis that adolescents are poor decision-makers, lacking in cognitive skills and understanding of consequences. More recent studies indicate that adolescents do not lack in perception or appraisal of risk, but that risk-taking is largely influenced by emotions and psychosocial factors, and cannot be completely understood in terms of cognitive processes during low emotions or arousal (Steinberg and Morris 2001; Steinberg 2004, 2005). This fits with the present findings, where participants believed they would benefit from taking precautions, yet finding motivation was difficult.

Two participants worried about developing active AIP, and it is worth noting that both had relatives with more chronic complaints. Perception of risk is subjective and dependent of context and the expected nature of potential outcome (Austin 2010). That disease presentation in the immediate family might influence perception of the condition has been suggested in previous research (Michie et al. 2002; Tønder et al. 2003). Family members attributing a variety of symptoms to AIP, and participants' vague understanding of latent and manifest forms emphasise the need for relevant and unambiguous information. For patients to benefit from predictive testing and early diagnosis, knowledge of genetic status is not enough. The ability to use risk information in a productive and constructive way is essential.

This implies that genetic counselling should be tailored to individual needs, providing sufficient information on symptoms, treatment, medical follow-ups and information on judicial rights (Paus 2009). Participants also indicated that need for information varied according to their situation in life and this favours the opportunity to access repeated information through, for example, patient education seminars, news-letters, web pages or individual counselling, and might be especially important when genetic testing is performed on minors.

That latent AIP triggered few lifestyle moderations could be viewed as positive in minimising the potential ethical dilemma of stigmatising healthy individuals. On the other hand, the goals of genetic service should ultimately be to improve long-term health status (Wang et al. 2004). Few consequences might question the direct benefit patients achieve through predictive testing as minors, as proposed by ethical guidelines (Borry et al. 2006).

In this study, it was positive to note that participants claimed to check medications for adverse porphyrinogenic reactions, and this is likely to prevent future AIP attacks (Badminton and Elder 2005). In addition, participants experienced early genetic testing as beneficial, and although there were reports of discrimination, this was not perceived as negative or dramatic. Minimal adverse response to testing has also been reported in previous research (Wang et al. 2005). Participants were also satisfied with their current knowledge and felt capable of obtaining further information when needed.

Strengths and Limitations

This study did not include persons who were not genetically tested or were tested as consenting adults. Reasons for, and consequences of, choosing to wait or not testing are therefore not examined. Qualitative findings can provide useful understanding and descriptions of the challenges minors tested for AIP experience, but the results cannot lend themselves to statistical analysis and transferability should be considered in relation to context and sociocultural setting (Kuper et al. 2008). Further investigations into psychosocial aspects and quality of life in persons with both latent and manifest AIP are subsequently desired.

Conclusion and Clinical Implications

AIP entailed relatively few psychosocial consequences for the participants, but results point to a potential for increased anxiety, without necessarily generating lifestyle changes. Discrimination based on genetic status was reported, and while the participant's experience of this was not negative, ethical implications need to be considered when predictive genetic testing is performed. Appropriate counselling and informed consent is advisable. What constitutes a suitable age for testing will differ from individual to individual, but these results suggest that parents in cooperation with their children may, after genetic counselling, be able to decide what age is appropriate.

References

Austin JC (2010) Re-conceptualising risk in genetic counseling: implications for clinical practice. J Genet Couns 19(3):228–234

Badminton MN, Elder GH (2005) Molecular mechanisms of dominant expression in porphyria. J Inherit Metab Dis 28:277–286

Borry P, Stultiens L, Cassiman NH, J-J DK (2006) Presymptomatic and predictive genetic testing in minors: a systematic review of guidelines and position papers. Clin Genet 70:374–381

Borry P, Goffin T, Nys H, Dierickx K (2008) Predictive genetic testing in minors for adult-onset genetic diseases. Mt Sinai J Med 75:287–296

Crimlisk HL (1997) The little imitator - porphyria: a neuropsychiatric disorder. J Neurol Neurosurg Psychiatr 62:319–328

Elder GH (1997) Hepatic porphyrias in children. J Inherit Metab Dis 20:237–246

Herrick AL, McColl KEL (2004) Acute intermittent porphyria. Best Pract Res Clin Gastroenterol 19(2):235–249

Hultin J, Schauman A, Wikberg A, Dahlquist G, Andersson C (2003) Acute intermittent porphyria in childhood: a population – based study. Acta Pædiatr 92:562–568

Kuper A, Lingard L, Levinson W (2008) Critically appraising qualitative research. BMJ 337:687–689

Michie S, French DP, Marteau T (2002) Predictive genetic testing: mediators and moderators of anxiety. Int J Behav Med 9:309–321

Millward LM, Kelly P, Deacon A, Senior V, Peters TJ (2001) Self-rated psychosocial consequences and quality of life in the acute porphyrias. J Inherit Metab Dis 24:733–747

Millward LM, Kelly P, King A, Peters TJ (2005) Anxiety and depression in the acute porphyrias. J Inherit Metab Dis 28:1099–1107

Morse J (2008) Confusing categories and themes. Qual Health Res 18(6):727–728

Normann Y, Puy H (2002) Human hereditary hepatic porphyrias. Clin Chim Acta 325(1–2):17–37

Paus B (2009) Klinisk genetikk: en innføringsbok. Gyldendal akademisk, Oslo

Pronk NP, Anderson LH, Crain AL et al (2004) Meeting recommendations for multiple healthy lifestyle factors: prevalence, clustering, and predictors among adolescent, adult, and senior health plan members. Am J Prev Med 27:25–33

Sandberg S, Brun A, Skadberg Ø, Iversen IR, Benestad Y, Danielsen OK (2001) Acute intermittent porphyria is a difficult diagnosis – especially in children. Tidsskr Nor Lægeforen 121:2822–2825

Sassa S (2006) Modern diagnosis and management of the porphyrias. Br J Haematol 135:281–292

Shiloh S, Ilan S (2005) To test or not to test? Moderators of the relationship between risk perception and interest in predictive genetic testing. J Behav Med 28:467–479

Steinberg L (2004) Risk taking in adolescence. What changes, and why? Ann N Y Acad Sci 1021:51–58

Steinberg L (2005) Cognitive and affective development in adolescence. Trends Cogn Sci 9(2):69–74

Steinberg L, Morris AS (2001) Adolescent development. Annu Rev Psychol 52:83–110

Thorne S, Kirkham SR, MacDonald-Emes J (1997) Interpretive description: a noncategorical qualitative alternative for developing nursing knowledge. Res Nurs Health 20:169–177

Thorne S, Kirkham R, O'Flynn-Magee K (2004) The analytic challenge in interpretive description. Int J Qual Method 3(1): Article 1. http://www.ualberta.ca/~iiqm/backissues/3_1/html/thorneetal.html

Tønder SL, Raaheim M, Skadberg Ø, Sandberg S (2003) Being prepared, but prepared for what? A qualitative study on the experience of genetic counselling and living with a predisposition for acute intermittent porphyria. Abstract presented at the International Porpyrins and Porphyrias Conference, Prague, September 21–24

Wade CH, Wilfond BS, McBride CM (2010) Effects of genetic risk information on children's psychosocial wellbeing: a systematic review of the literature. Genet Med 12(6):317–326

Wang C, Gonzalez R, Merajver S (2004) Assesment of genetic testing and related counseling services: current research and future directions. Soc Sci Med 58:1427–1442

Wang C, Bowen DJ, Kardia SLR (2005) Research and practice opportunities at the intersection of health education, health behavior, and genomics. Health Educ Behav 32:686–701

Wikberg A, Jansson L, Lithner F (2000) Women's experience of suffering repeated severe attacks of acute intermittent porphyria. J Adv Nurs 32(6):1348–1355

JIMD Reports
DOI: 10.1007/8904_2011_9

Received: 30 August 2010; Revised: 26 October 2010;
Accepted: 30 December 2010; Published online: 22 June
Communicated by: Frits Wijburg

Enzyme Replacement Therapy and Extended Newborn Screening for Mucopolysaccharidoses: Opinions of Treating Physicians

David J. Coman, Ian M. Hayes, Veronica Collins, Margaret Sahhar, J. Ed Wraith, and Martin B. Delatycki

Abstract We conducted a survey of physician opinions in relation to enzyme replacement therapy (ERT) and extended newborn screening (ENBS) for mucopolysaccharidoses (MPS). A questionnaire consisting of hypothetical clinical scenarios about ERT and ENBS for MPS was posted on metab-L, a list server for the metabolic community. The questionnaire included similar questions to those used in previous studies that sought the views of individuals and families affected by MPS. Our aim was to compare medical professionals' opinions with that of the individuals and families affected by MPS that they serve. The questionnaire was completed by 35 physicians, most of whom were metabolic physicians. Responses differed significantly between the physician and parent groups when the clinical scenario involved intellectual impairment. In this setting, physicians were significantly less inclined to advocate the use of ERT. Comparison of the responses to the ENBS scenarios revealed that compared to physicians, family of individuals with MPS were more inclined to desire diagnosis at birth, even if no treatment could alter the outcome of the condition. Compared to the family of individuals with MPS, physicians are more likely to advocate the use of ERT and ENBS where there is proven medical benefit to the affected individual.

Keywords Ethics · Lysosomal storage disorders · Questionnaire

D.J. Coman
Bruce Lefroy Centre for Genetic Health Research, Murdoch Childrens Research Institute, 10th Floor, Royal Children's Hospital, Flemington Road, Parkville 3052, Victoria, Australia
and
Department of Metabolic Medicine, The Royal Children's Hospital, Brisbane Queensland, Australia

I.M. Hayes
Bruce Lefroy Centre for Genetic Health Research, Murdoch Childrens Research Institute, 10th Floor, Royal Children's Hospital, Flemington Road, Parkville 3052, Victoria, Australia
and
Northern Regional Genetic Service, Auckland City Hospital, Auckland New Zealand

V. Collins
Bruce Lefroy Centre for Genetic Health Research, Murdoch Childrens Research Institute, 10th Floor, Royal Children's Hospital, Flemington Road, Parkville 3052, Victoria, Australia
and
Public Health Genetics Unit, Murdoch Childrens Research Institute, Parkville Victoria, Australia

M. Sahhar
Bruce Lefroy Centre for Genetic Health Research, Murdoch Childrens Research Institute, 10th Floor, Royal Children's Hospital, Flemington Road, Parkville 3052, Victoria, Australia
and
Genetic Health Services Victoria, Parkville Victoria Australia

J.E. Wraith
Department of Genetic Medicine, St. Mary's Hospital, Manchester UK

M.B. Delatycki (✉)
Bruce Lefroy Centre for Genetic Health Research, Murdoch Childrens Research Institute, 10th Floor, Royal Children's Hospital, Flemington Road, Parkville 3052, Victoria, Australia
and
Department of Clinical Genetics, Austin Health, Heidelberg Victoria, Australia
and
Department of Medicine, University of Melbourne, Austin Health, Heidelberg Victoria, Australia
e-mail: martin.delatycki@ghsv.org.au

Introduction

The mucopolysaccharidoses (MPS) are a group of related disorders of lysosomal storage that have an approximate prevalence of 1:22,500 live births (Meikle et al. 1999). Diminished levels of specific lysosomal enzymes result in progressive accumulation of glycosaminoglycans (GAGs) in the lysosomes of various organs. The resultant clinical manifestations can include intellectual disability, hydrocephalus, spinal cord compression, deafness, dysostosis multiplex, short stature, hepatosplenomegaly, sleep apnoea, cardiac disease, corneal clouding, coarse facial features, and reduced life span in the majority (McKusick 2001).

Early bone marrow transplantation is effective in the severe form of MPSI (MPS-1H) (Pastores and Barnett 2005). Enzyme replacement therapy (ERT) can ameliorate some clinical features associated with select subtypes of MPS and is now available for MPSI (Laronidase), MPSII (Idursulfase), and MPSVI (Galsulfase) (Muenzer et al. 2006; Kakkis 2002; Harmatz et al. 2006). Significant limitations of ERT include that ERT will not cross the blood–brain barrier and thus do not impact the central nervous system manifestations of MPS and the very high cost, and resultant limited drug availability in some countries.

Newborn screening (NBS) methods have been developed for the identification of individuals at birth who have specific lysosomal storage disorders (Meikle et al. 2006; Wang et al. 2005, 2007). Glycosphingolipid and oligosaccharide markers (measured using tandem mass spectroscopy) have 100% sensitivity and specificity for the identification of MPSIIIA, MPSIVA, I-cell disease as well as a number of other lysosomal storage disorders (Janssen et al. 2005). Measurement of specific enzymes deficient in other types of MPS would enable these to be identified by NBS (Wang et al. 2007; De Jesus et al. 2009).

We have previously conducted questionnaire studies to assess the opinions of parents of individuals with MPS and adults with MPS in the USA and Australia regarding NBS and ERT for these conditions (Coman et al. 2008; Hayes et al. 2007). Eighty-six percent of respondents indicated that they would have wanted NBS for their own children, with the most common reason cited in support of NBS being the avoidance of a delay in diagnosis and the distress caused by a delayed diagnosis (Hayes et al. 2007). This study indicated strong support for the introduction of NBS from MPS families, in which the psychosocial benefits of screening may outweigh potential harms. Overall, 92% were in favor of ERT where MPS causes severe physical problems but does not affect intellect, and 69% were in favor of ERT where the physical limitations are mild and intellect is spared (Coman et al. 2008). The majority of respondents were in favor of ERT for MPS, even where it would not alter the intellectual deterioration, with perceived improvements in quality of life (QOL) produced by ERT, the most commonly cited factor (Coman et al. 2008).

In considering treatments such as ERT, the clinician must consider factors such as (1) the prospects for benefit, (2) parental wishes, (3) potential negative impacts on the patient and the wider family unit, and (4) the cost of treatment. We undertook a questionnaire study to elicit the views of physicians who treat MPS, on ERT and NBS for MPS. We hypothesized that clinicians would have differing views to the MPS families regarding the suitability of NBS and ERT in some MPS patient scenarios, especially where intellectual impairment is involved in the clinical phenotype.

Methods

The questionnaire used in this study was based on that used in our previous studies to elicit the views of individuals with MPS and parents (Coman et al. 2008; Hayes et al. 2007) but included some questions specifically designed for medical professionals. It included questions on the clinical role of the respondent, how many patients they have treated with each MPS subtype, and their responses to a number of hypothetical clinical scenarios regarding ERT and extended newborn screening (ENBS) for MPS. The questionnaire with details of these scenarios is available at http://www.mcri.edu.au/Downloads/Survey/ERT_expert_survey.pdf, and the scenarios are summarized in Tables 1 and 2. Questionnaires were distributed online via metab-L (http://www.daneel.franken.de/metab-l), a list server for the worldwide metabolic community with over 1,000 active members. The metab-L post-directed respondents to the MCRI website where the questionnaire was completed anonymously.

Statistical Analysis

Characteristics of the health professionals who responded to the survey are provided as frequency data. Physician responses are presented as the proportion of respondents agreeing with the use of ERT or ENBS for each scenario. The responses of the physicians were compared with those of the individuals affected by MPS and parents, elicited in earlier studies (Coman et al. 2008; Hayes et al. 2007), using chi-square statistics with one degree of freedom or Fisher's exact tests where an expected cell frequency was less than five. The level of evidence for an association between responses to scenarios and whether the respondent was a physician or an affected individual/parent is given as a p value. A p value of < 0.05 was considered statistically significant.

Results

Thirty-five physicians responded to the survey. Metabolic physicians were the highest represented specialty among the respondents (65.7%) followed by medical geneticists (11.4%), general pediatricians (8.5%), with three respondents not stating their specialty, and two indicating another specialty

Table 1 Outline of the six ERT scenarios presented in the questionnaire

	Scenario summary	Manifestations		Effect of ERT		
		Physical	Intellectual	Physical	Life expectancy	Intellectual
1a	11-year old with severe physical MPSVI and normal intellect, ERT will prolong life	Severe	Nil	Significant improvement	Increased	Nil improvement
2a	MPSI-S with minor health issues as an adult but normal intellect, ERT will only improve health issues a little	Mild	Nil	Minimal improvement	Unchanged	Nil improvement
3a	Severe physical and intellectual manifestations of MPSII. Death in 5 years without ERT	Severe	Severe	Minimal improvement	Increased 3 years	Nil improvement
4a	The 1-year-old younger brother of the MPSII patient in 3a. ERT will prolong his life but not halt the CNS manifestations	Moderate	Severe	Minimal improvement	Increased 15 years	Nil improvement
5a	Mild MPSII with early joint and mobility difficulties, and mild school performance issues	Mild	Mild	Moderate improvement	Not significant	Nil improvement
3b	–[a]	Severe	Severe	Minimal improvement	Increased 3 years	Significant improvement
4b	–[a]	Moderate	Severe	Minimal improvement	Increased 15 years	Significant improvement
5b	–[a]	Mild	Mild	Moderate improvement	Unchanged	Significant improvement

[a]3b, 4b, and 5b are identical to 3a, 4a, and 5a; however, the respondents were asked to consider whether they would use ERT for the patients in 3a, 4a, and 5a if it improved the CNS manifestations

Table 2 Outline of three NBS scenarios presented in the questionnaire

	MPS type	Manifestations	
		Physical	Intellectual
Scenario 1	MPSVI	Mild	Nil
Scenario 2	MPSIII	Mild	Severe
Scenario 3	MPSI	Severe	Severe

Table 3 Characteristics of the participating physicians ($n = 35$)

	Number	Proportion (%)
Specialty		
Metabolic physician	23	65.7
Medical geneticist	4	11.4
Pediatrician	3	8.6
Other	2	5.7
Not stated	3	8.6
Years in practice		
0–10	15	42.8
11–20	8	22.9
More than 20	8	22.9
Not stated	4	11.4

(Table 3). Under half (42.8%) of respondents had been practicing in the field for 10 years or less with 22.8% very experienced as evidenced by working in the field for over 20 years.

The numbers of past and present MPS I, II, and VI patients treated by the responding physicians are outlined in Table 4, indicating a broad range of experience with these conditions. Unsurprisingly, given its novelty and expense, there was relatively limited experience with the use of ERT (Table 5).

Enzyme Replacement Therapy

The hypothetical ERT scenarios are outlined in Table 1, and the physician responses are shown in Table 6 where they are compared to those of the individuals with MPS/families (Coman et al. 2008).

Responses of the physician and parent groups followed certain themes. Both groups were united in their opinions that ERT should be used in situations where intellect is normal with only mild physical problems (scenario 1a), mild decreases in IQ with mild joint restriction improved by ERT (scenario 5a), severe intellectual disability where ERT would prolong life by 15 years (scenario 4a), and in hypothetical scenarios where ERT can avert the evolving CNS manifestations (scenarios 4b and 5b).

However, differences were observed in other scenarios. Physicians were significantly less inclined to support the use of ERT in scenarios where severe intellectual impairment is prominent (scenario 3a: 8.8% physicians in favor vs. 46.9% families in favor, $p < 0.001$; scenario 3b: 38.2% physicians in favor vs. 82.6% families in favor, $p < 0.001$) or when the child displays only mild physical manifestations (scenario 2a: 38.2% physicians in favor vs. 69.2% families in favor, $p < 0.001$).

Forty percent (14/35) of the physician group made additional comments. The most common themes among the comments included the need to involve the affected individual/families in the ERT decision process (4/14, 28.5%), concerns that ERT may prolong suffering rather than prolong

Table 4 Number and type of MPS patients treated by physicians in the study (current and past)

MPS type	Current care			Past care		
	Number responded	Median number of patients	Total number of patients	Number responded	Median number of patients	Total number of patients
MPSI	30	5.7	171	26	8.5	220
MPSII	26	5.7	150	23	6	139
MPSIII	22	8.7	193	21	8.6	180
MPSIV	21	4.2	90	18	5.2	94
MPSVI	20	3.3	67	16	4.3	69
MPSVII	9	0.8	8	6	0.66	4

Table 5 Number and type of MPS patients treated by ERT

MPS type	ERT patient numbers				
	Nil	1–10	11–20	21–30	No response
MPSI	9 (25.7%)	20 (57%)	0 (0%)	1 (2.8%)	5 (14.3%)
MPSII	7 (20%)	21 (60%)	0 (0%)	1 (2.8%)	6 (17%)
MPSVI	6 (17%)	18 (51.4%)			11 (31.4%)

MPSIII, MPSIV, and MPSVII not applicable as no ERT is currently available for these subtypes of MPS

Table 6 Comparison of proportions of respondents who agreed with the use of ERT in each scenario between physicians and patients/parents

Scenario	Physicians		Patients/parents[a]		p value
	No. responded	% agree with ERT	No. responded	% agree with ERT	
1a	34	91.2	245	91.8	0.896
2a	34	38.2	247	69.2	**<0.001**
3a	34	8.8	245	46.9	**<0.001**
4a	34	67.6	241	76.8	0.246
5a	33	90.9	247	92.7	0.712
3b	34	38.2	242	82.6	**<0.001**
4b	34	100	243	97.1	0.603[b]
5b	34	97.1	241	98.3	0.486[b]

[a]Affected individual/parent responses from previous study (Coman et al. 2008)
[b]Fisher's exact test

QOL in some MPS scenarios (4/14, 28.5%), and that the use of ERT is not ideal in situations of intellectual impairment (4/14, 28.5%). The most common additional comments cited in favor of ERT in the family study included improved QOL (26% of those who made additional comments), that ERT should be available to all regardless of intellectual status (18%), with 4% of parents of children with MPS stating that ERT should not be used where it would prolong suffering (Coman et al. 2008).

Expanded Newborn Screening

The hypothetical NBS scenarios are outlined in Table 2, and the physician responses are listed in Table 7 where they are compared to those of the individuals with MPS/families (Hayes et al. 2007).

Table 7 Comparison of proportions of respondents who agreed with having NBS in each scenario between physicians and patients/parents

Scenario	Physicians		Patients/parents[a]		p value
	No. responded	% agree with ENBS	No. responded	% agree with ENBS	
1	34	64.7	242	83.9	**<0.01**
2	34	64.7	247	87.4	**<0.001**
3	35	100	245	97.1	0.60[b]

[a]Affected individual/parent responses from previous study (Hayes et al. 2007)
[b]Fisher's exact test

There was no significant difference between the groups in relation to the response to scenario 3 where NBS would allow early diagnosis of MPSI such that bone marrow transplantation can be offered (100% of physicians in favor and 97.1% of families in favor, $p = 0.60$). There was a significantly lower positive response by the physicians compared to individuals with MPS/families for scenario 1 (mild case of MPSVI where ERT would not be required), where NBS was advocated by 64.7% of the physicians and 83.9% of the affected individuals/families ($p = 0.007$); and scenario 2 (MPSIII) where NBS was advocated by 64.7% of physicians and 87.4% of the affected individuals/families ($p < 0.001$).

Overall, 42.8% (15/35) of physician respondents made additional comments. The most frequently cited reasons in favor of NBS were the ability to provide genetic counseling to affected families to allow the option of preventing the birth of further children with the condition (7/15 of those who made comments, 46.6%), and health benefits and treatments potentially made available by early diagnosis (7/15, 46.6%). The most frequent comments in favor of NBS in the family study were increased stress associated with not having a diagnosis (25%), improved medical care (24%), and genetic counseling (19%) (Hayes et al. 2007). The most commonly cited arguments against NBS cited by physicians in this study were the inability to judge phenotype from the NBS results and thus predict those in need of treatment where available (4/14, 26.6%). Only one clinician made comments about the possible impact on families from false-negative and false-positive NBS results. Themes of potential harm were cited in the family study, with NBS removing a period of perceived good health (7%) and creating an altered and negative perception of their child (2%).

Discussion

MPS presents with a wide range of clinical manifestations which can create significant morbidity and mortality. The emergence of ERT has been welcomed by clinicians and families; however, it carries significant limitations. Clinical trials of ERT in MPS I, II, and VI have demonstrated reduced disease burden for many non-CNS manifestations, but ERT does not impact on CNS morbidity (Muenzer et al. 2006; Harmatz et al. 2006; Wraith et al. 2004; Kakkis et al. 2001).

We previously found that MPS families have a sound understanding of the uses and limitations of ERT for MPS (Coman et al. 2008). Nevertheless, we hypothesized that variation would exist in how medical professionals and families judge burden of disease, especially where intellectual impairment is involved, and thus the appropriateness of ERT in these situations. Physicians were significantly less inclined than parents to think that ERT was an appropriate therapeutic option in scenarios where severe intellectual impairment is prominent (scenario 3a and 3b) or when the child displays only mild physical manifestations (scenario 2a). Further evidence for differing views on the appropriateness of ERT in these situations came from the additional comments. A commonly cited theme from the physicians was concern regarding the use of ERT where life might be prolonged without associated improvement in QOL, where intellectual impairment was involved.

MPS creates a significant burden of disease for the individual, the family, and health care systems, which can be judged in terms of the clinical manifestations (intellectual and physical), functional limitations, familial dysfunction and dislocation, invasive medical intervention (especially weekly ERT), impacts on hospital resources, and monetary cost to society (again exacerbated by ERT). QOL is defined as the individual's perception of their physical and psychological health, level of independence, and social and personal relationships that are tempered by the environment and culture in which the individual lives (Saxena and Orley 1997). Our study demonstrates affected individual/family-physician discordance regarding the use of ERT when the disease burden involves intellectual impairment. Physicians have been shown to misjudge the perceived QOL in patients with intellectual disability (Janssen et al. 2005), and underestimate the patient's positive view of their own QOL (Nursey et al. 1990). Similar discordance in assessing disease severity, management decisions, and perceived QOL has been observed in rheumatoid disorders (Consolaro et al. 2007; Spoorenberg et al. 2005; Yen et al. 2003), planning gastroenterological procedures (Sonnenberg 2004), inflammatory bowel disease (Sewitch et al. 2002), schizophrenia (Fitzgerald et al. 2001), and malignancy (Wilson et al. 2000). ERT is very expensive in its own right, but its administration also creates collateral costs on health professional's time and resources beyond the time pressures on the individual and family.

Affected individual/families may well be able to make a valid argument that the child's non-CNS health and QOL would be improved with ERT, and not providing ERT to their child is discrimination on the basis of symptoms. Ultimate decisions regarding the appropriateness of ERT for individuals with progressive intellectual deterioration need to be made in conjunction with government bodies, treating physicians and other health professionals and families. An example of this process is the UK guidelines for investigations and management of MPSI (http://www.dh.gov.uk/en/Publicationsandstatistics/Publications/PublicationsPolicyAndGuidance/DH_4118402).

Both physicians and family groups were supportive of NBS where the early diagnosis would lead to a proven therapeutic option, e.g., BMT for MPSI (scenario 3). Highly statistically significant variation was observed, however, for the use of NBS in a scenario where it would diagnose mild MPSVI that would not qualify for ERT (scenario 1) and for MPSIII (scenario 2), where no treatment can prevent the inexorable decline of this condition. MPSIII has a severe CNS phenotype including intellectual impairment and regression but far fewer non-CNS manifestations than other MPSs. The extra comments from families suggest that it is psychosocial rather than medical benefits that underlie this difference. The themes emerging from additional comments support this contention, especially revolving around the stress of a delayed diagnosis (Hayes et al. 2007).

A significant limitation of NBS at present is the inability to accurately predict phenotype and projected treatment requirements, something noted by the physician group, highlighting the need for ongoing education of affected individuals and their families. False-positive results would undoubtedly cause significant stress to families, and proven diagnoses may "rob" the family of a period of normality when no treatment options are available to alter later onset disease progression, and may even create a negative impact on parent–child bonding. However, the latter concern has not been the reported experience in Duchene muscular dystrophy newborn screening in Wales, which uses detailed parent–physician communication strategies (Parsons et al. 2002). These are issues that were raised by the affected individual/family group but not the physicians (Hayes et al. 2007). Guidelines state that NBS should only be introduced if a disorder is severe, and early diagnosis results in a better outcome for the affected individual (Wilcken 2008). Early diagnosis afforded by NBS might allow timely genetic counseling and family planning, and was raised as a positive outcome by both physicians and affected individuals/ families as an important potential benefit of NBS.

This study has some limitations. Ascertainment bias may exist by virtue of the small sample of clinicians who responded to the survey, with varying degrees of experience with the use of ERT, and therefore the range of responses to the hypothetical scenarios may be biased in those motivated to complete the online survey. Thus, the views of the physicians who responded to the survey may not reflect those of the wider medical community involved in treating MPS. Continental or cultural variances and the impacts of the respondent's area of medical specialization on physician's opinions were not explored due to the small sample size. The results of the current study were compared to the results of previous studies that assessed the responses of patients/parents to the same scenarios (Coman et al. 2008; Hayes et al. 2007). The time interval between the two studies may have resulted in bias as the patients/parents may have responded differently now, since experience with ERT has considerably increased during the time since the initial survey was conducted.

The advent of ERT and the evolution of NBS create many ethical and sociological issues that impinge on the physician–family interface. Clinicians face the difficult dual task of advocating for their patients and developing guidelines for the use of expensive medications that impact on already stretched health budgets and systems. Involvement of advocacy groups in formulating these policies offers an opportunity for informing both parties as to the range of opinions about appropriateness for treatment, as indicated in this study. Failure to involve families in such important matters may alienate the treating clinicians and funding bodies from those that they advocate for and may even be construed as paternalistic.

Acknowledgments We thank the MPS societies of Australia and the USA for distributing the questionnaires and those who took the time to complete the surveys. We thank Drs. Heidi Peters and Joy Lee for their helpful comments in relation to the manuscript. We also thank the Information Technology Department at Murdoch Childrens Research Institute for their assistance in developing the web-based questionnaire. MBD is an NHMRC Practitioner Fellow. All authors have contributed to the planning, data interpretation, and manuscript production, and have reviewed the material, agreed to its content, and declared no conflict of interests. Ethics approval and funding were not required for this survey-based study. Prof. Martin B Delatycki is the guarantor for the manuscript.

Take Home Message

Compared to the family of individuals with MPS, physicians are more likely to advocate the use of enzyme replacement therapy and extended newborn screening where there is proven medical benefit to the affected individual.

References

Coman DJ, Hayes IM, Collins V, Sahhar M, Wraith JE, Delatycki MB (2008) Enzyme replacement therapy for mucopolysaccharidoses: opinions of patients and families. J Pediatr 152:723–727

Consolaro A, Vitale R, Pistorio A, Lattanzi B, Ruperto N, Malattia C et al (2007) Physicians' and parents' ratings of inactive disease are frequently discordant in juvenile idiopathic arthritis. J Rheumatol 34:1773–1776

De Jesus VR, Zhang XK, Keutzer J, Bodamer OA, Muhl A, Orsini JJ et al (2009) Development and evaluation of quality control dried blood spot materials in newborn screening for lysosomal storage disorders. Clin Chem 55:158–164

Fitzgerald PB, Williams CL, Corteling N, Filia SL, Brewer K, Adams A et al (2001) Subject and observer-rated quality of life in schizophrenia. Acta Psychiatr Scand 103:387–392

Harmatz P, Giugliani R, Schwartz I, Guffon N, Teles EL, Miranda MC et al (2006) Enzyme replacement therapy for mucopolysaccharidosis VI: a phase 3, randomized, double-blind, placebo-controlled, multinational study of recombinant human N-acetylgalactosamine 4-sulfatase (recombinant human arylsulfatase B or rhASB) and follow-on, open-label extension study. J Pediatr 148:533–539

Hayes IM, Collins V, Sahhar M, Wraith JE, Delatycki MB (2007) Newborn screening for mucopolysaccharidoses: opinions of patients and their families. Clin Genet 71:446–450

Janssen CG, Schuengel C, Stolk J (2005) Perspectives on quality of life of people with intellectual disabilities: the interpretation of discrepancies between clients and caregivers. Qual Life Res 14:57–69

Kakkis ED (2002) Enzyme replacement therapy for the mucopolysaccharide storage disorders. Expert Opin Investig Drugs 11:675–685

Kakkis ED, Muenzer J, Tiller GE, Waber L, Belmont J, Passage M et al (2001) Enzyme-replacement therapy in mucopolysaccharidosis I. N Engl J Med 344:182–188

McKusick VA (2001) The mucopolysaccharidoses, 8th edn. McGraw-Hill, New York. J Inherit Metab Dis 31:173–177

Meikle PJ, Hopwood JJ, Clague AE, Carey WF (1999) Prevalence of lysosomal storage disorders. JAMA 281:249–254

Meikle PJ, Grasby DJ, Dean CJ, Lang DL, Bockmann M, Whittle AM et al (2006) Newborn screening for lysosomal storage disorders. Mol Genet Metab 88:307–314

Muenzer J, Wraith JE, Beck M, Giugliani R, Harmatz P, Eng CM et al (2006) A phase II/III clinical study of enzyme replacement therapy with idursulfase in mucopolysaccharidosis II (Hunter syndrome). Genet Med 8:465–473

Nursey AD, Rohde JR, Farmer RD (1990) A study of doctors' and parents' attitudes to people with mental handicaps. J Ment Defic Res 34(Pt 2):143–155

Parsons EP, Clarke AJ, Hood K, Lycett E, Bradley DM (2002) Newborn screening for Duchenne muscular dystrophy: a psychosocial study. Arch Dis Child Fetal Neonatal Ed 86:F91–F95

Pastores GM, Barnett NL (2005) Current and emerging therapies for the lysosomal storage disorders. Expert Opin Emerg Drugs 10:891–902

Saxena S, Orley J (1997) Quality of life assessment: the world health organization perspective. Eur Psychiatry 12(Suppl 3):263s–266s

Sewitch MJ, Abrahamowicz M, Bitton A, Daly D, Wild GE, Cohen A et al (2002) Psychosocial correlates of patient-physician discordance in inflammatory bowel disease. Am J Gastroenterol 97:2174–2183

Sonnenberg A (2004) Patient-physician discordance about benefits and risks in gastroenterology decision-making. Aliment Pharmacol Ther 19:1247–1253

Spoorenberg A, van Tubergen A, Landewe R, Dougados M, van der Linden S, Mielants H et al (2005) Measuring disease activity in ankylosing spondylitis: patient and physician have different perspectives. Rheumatology (Oxford) 44:789–795

Wang D, Eadala B, Sadilek M, Chamoles NA, Turecek F, Scott CR et al (2005) Tandem mass spectrometric analysis of dried blood spots for screening of mucopolysaccharidosis I in newborns. Clin Chem 51:898–900

Wang D, Wood T, Sadilek M, Scott CR, Turecek F, Gelb MH (2007) Tandem mass spectrometry for the direct assay of enzymes in dried blood spots: application to newborn screening for mucopolysaccharidosis II (Hunter disease). Clin Chem 53: 137–140

Wilcken B (2008) The consequences of extended newborn screening programmes: do we know who needs treatment? J Inherit Metab Dis 2008

Wilson KA, Dowling AJ, Abdolell M, Tannock IF (2000) Perception of quality of life by patients, partners and treating physicians. Qual Life Res 9:1041–1052

Wraith JE, Clarke LA, Beck M, Kolodny EH, Pastores GM, Muenzer J et al (2004) Enzyme replacement therapy for mucopolysaccharidosis I: a randomized, double-blinded, placebo-controlled, multinational study of recombinant human alpha-L-iduronidase (laronidase). J Pediatr 144:581–588

Yen JC, Abrahamowicz M, Dobkin PL, Clarke AE, Battista RN, Fortin PR (2003) Determinants of discordance between patients and physicians in their assessment of lupus disease activity. J Rheumatol 30:1967–1976

JIMD Reports
DOI: 10.1007/8904_2011_10

Received: 4 October 2010; Revised: 28 October 2010;
Accepted: 15 December 2010; Published online: 22 June
Communicated by: Verena Peters

Mutation Spectrum of Fumarylacetoacetase Gene and Clinical Aspects of Tyrosinemia Type I Disease

A. Dursun, R.K. Özgül, S. Sivri, A. Tokatlı, A. Güzel, L. Mesci, M. Kılıç, D. Aliefendioglu, F. Özçay, M. Gündüz, and T. Coşkun

Abstract Tyrosinemia type I (OMIM 276700) is a rare, autosomal recessive disorder caused by a deficiency in the fumarylacetoacetate hydrolase (FAH) enzyme. This study examined the spectrum of *FAH* gene mutation in 32 patients with tyrosinemia type I. In addition, clinical and biochemical findings were evaluated to establish a genotype–phenotype relationship in the patients. Mutation screening was performed using a 50K custom-designed resequencing microarray chip (TR_06_01r520489, Affymetrix) and sequencing analysis. Of the 12 different mutations found, 6 are categorized as novel. Three of the mutations-IVS6-1G>A, D233V, and IVS3-3C>G-are the most common in Turkish patients, comprising 25%, 17.1%, and 12.5% of mutant alleles, respectively.

Clinical evaluations suggest that the spectrum of symptoms observed in the patients with very early and early disease were of the more nonspecific form, whereas the patients with late-presenting disease had more of the distinctive form over the course of the disease. This study adds support to the notion that the D233V mutation is specific to the Turkish population.

Keywords FAH mutations · Genotype · Microarray · Phenotype · Resequencing chip · Tyrosinemia type I

A. Dursun (✉), S. Sivri, A. Tokatlı, L. Mesci, M. Kılıç, and T. Coşkun
Department of Pediatrics, Metabolism Unit, Hacettepe University, Ankara, Turkey
e-mail: adursun@hacettepe.edu.tr

R. Özgül
Department of Pediatrics, Metabolism Unit, Hacettepe University, Ankara, Turkey
and
Institute of Child Health, Hacettepe University, Ankara, Turkey

A. Güzel
Department of Pediatrics, Metabolism Unit, Hacettepe University, Ankara, Turkey
and
Department of Biology, Molecular Biology Section, Hacettepe University, Ankara, Turkey

D. Aliefendioglu
Department of Pediatrics, Kırıkkale University, Kırıkkale, Turkey

F. Özçay
Department of Pediatrics, Başkent University, Ankara, Turkey

M. Gündüz
Ministry of Health, Dışkapı Research and Training Hospital, Ankara, Turkey

Introduction

Tyrosinemia type I (OMIM 276700) is an autosomal recessive disorder caused by a deficiency in the fumarylacetoacetate hydrolase (FAH) enzyme, which plays a role in the final step of tyrosine amino acid catabolism. The disease has panethnic distribution and varying frequencies worldwide. The disease primarily affects the liver and proximal renal tubular system. Liver disease can manifest as acute hepatic failure, mixed micro nodular cirrhosis, hepatocellular dysfunction, and hepatocellular carcinoma, and proximal tubular dysfunction causes such symptoms as hypophosphatemic rickets accompanying aminoaciduria, renal tubular acidosis, and glucosuria. Acute neurological crisis and cardiomyopathy may also develop in patients with the disease.

Three clinical phenotypes of the disease are described based on age at onset of symptoms: the acute form presents before 6 months of age with acute liver failure, the subacute form presents between 6 months and 1 year of age with liver disease, and the chronic form presents after 1 year of age with slowly progressive liver cirrhosis and hypophosphatemic rickets (Bergman et al. 1998; Mitchell et al. 2001; Chakrapani and Holme 2006). Van Spronsen et al. (1994, 1995) suggested another classification for tyrosinemia type I on the basis of survival rates and the age of symptom onset: very early form (onset of symptoms at <2 months of age), early form (onset of symptoms at 2–6 months of age), and late-presenting form (onset of symptoms at >6 months of age). Laboratory findings for tyrosinemia type I include increased plasma levels of tyrosine (TYR), methionine (MET), and alpha-fetoprotein (AFP), and excessive urinary excretion of maleylacetoacetate (MAA), fumarylacetoacetate (FAA), and their derivatives-succinyl acetone (SA) and succinyl acetoacetate (SAA) (Chakrapani and Holme 2006).

The *FAH* gene has been mapped to chromosome 15q23-25 and contains 14 exons that code 420 amino acids (Awata et al. 1994). To date, more than 40 pathogenic mutations in the *FAH* gene have been reported in different populations. Some of the mutations appear to have accumulated in particular ethnic groups, i.e., IVS12+5G>A in French–Canadians (approximately 90% of alleles), IVS6-1G>T in the Mediterranean region (approximately 60% of alleles), W262X in Finns, D233V in Turks, and Q64H in Pakistanis (Rootwelt et al. 1996; St-Louis and Tanguay 1997; Bergman et al. 1998; Arranz et al. 2002; Elpeleg et al. 2002; Heath et al. 2002). Despite the fact that the spectrum of *FAH* gene mutation has been expanded, current knowledge is not adequate for establishing the disease's genotype–phenotype relationship.

This study examined the spectrum of *FAH* gene mutation in 32 patients with tyrosinemia type I. Of the 12 different mutations detected, 6 are categorized herein as novel. Based on the clinical and laboratory findings at the time of diagnosis, an attempt was made to establish a genotype–phenotype relationship.

Materials and Methods

In all, 32 Turkish patients with tyrosinemia type I that, to the best of our knowledge, came from unrelated families were examined for *FAH* gene mutations. Clinical diagnosis was based on increased levels of tyrosine in serum and the presence of SA in urine. No enzymatic studies were performed. The patients were categorized as very early (onset of symptoms at <2 months of age), early (onset of symptoms at 2–6 months of age), and late-presenting (onset of symptoms at >6 months of age), according to van Spronsen classification. Genotype and phenotype profiles, together with biochemical findings, are summarized in Table 1.

Mutation screening was performed using a 50K resequencing microarray chip (TR_06_01r520489/Affymetrix) custom designed by our research group to sequence all exonic sequences and their flanking intronic sequences for the following 14 genes responsible for 10 different types of inborn errors of metabolic diseases: the *ALDOB* gene for hereditary fructose intolerance, the *ATP7B* gene for Wilson disease, *BCKDHA*, *BCKDHB*, *DBT*, and *DLD* genes for maple syrup urine disease, the *FAH* gene for tyrosinemia type I, the *FBP1* gene for fructose 1–6 diphosphatase deficiency, the *GALT* gene for galactosemia, the *GCDH* gene for glutaric aciduria type I, the *MUT* gene for methylmalonic acidemia, *PCCA* and *PCCB* genes for propionic acidemia, and the *PAH* gene for phenylketonuria. Genomic DNA was extracted from blood samples using the salting-out technique. Fifty healthy individuals were selected as a control group to test the identified nucleotide changes and their frequencies in normal people. The study protocol was approved by the Hacettepe University Ethics Committee.

In brief, the *FAH* gene was amplified from genomic DNA via long-range polymerase chain reaction (PCR). The primer sets used for long-range PCR, and the related exons and amplicon sizes are given in Table 2. After purification (Qiagen kit), all PCR products were quantified (NanoDrop Technologies) and equimolar quantities were pooled. After the fragmentation step, fragmented PCR products were end-labeled using a biotin-labeling reagent and hybridized with DNA arrays (GeneChip Resequencing Assay Kit, Affymetrix). Arrays were processed via washing and staining on a fluidics station. Scanned arrays were analyzed using Affymetrix GeneChip resequencing analysis software.

Direct DNA sequencing was performed using primers specific for each exon to confirm all of the nucleotide changes detected by the microarray resequencing chip. Sequences of the exonic primers will be supplied upon request. For the sequencing reaction, BigDye Terminator Cycle Sequencing v.3.1 (Applied Biosystems, Foster City, CA) and sense or anti-sense primers were used in both the forward and reverse directions. An ABI 3130 capillary electrophoresis system was used for automated sequencing (Applied Biosystems) with the POP7 polymer. Sequencing chromogram files were analyzed using sequencing analysis software.

Results and Discussion

In total, two patients had very early disease, 15 had early disease, and eight had late-presenting disease. Six patients who were diagnosed via selective newborn screening because of a positive family history were excluded from clinical classification, and one patient, whose clinical and biochemical findings were not available, was excluded from classification. While two patients with the very early form presented with gastrointestinal hemorrhagia and acute liver failure that mimicked other acute neonatal problems, 15 patients with the early form had broader clinical symptomatology, including acute hepatic insufficiency, sepsis, hepatomegaly, chronic diarrhea, and rickets. On the contrary, hepatomegaly and rickets were primarily observed in patients with the late-presenting form.

Neurologic crisis and restrictive cardiomyopathy, which disappeared after the initiation of treatment, were noted during the follow-up of two patients with the late-presenting form. All but one patient excreted SA in their urine at

Table 1 Clinical and biochemical findings and genotypes of the patients

Patient No	Age of Diag. (mo)	Hepatic and tubular function									Mutation			
		Presenting symptoms	Tyr levels (mg/dl)	Synthetic function (PT/PTT)	ALT/AST	Total Bil. levels (mg/dl)	a-FP IU/ml[a] (N 0–5.8)	SA	Tubulopathy	Type	Exon	Nucleotide	Protein	Outcome
1	2 day	GIS bleeding	25.1	Abnormal	40/102	3	855.000	+	NA	VE	2	c.191delA (Hom)		Well
2	3	HM	10.6	Abnormal	49/83	1.1	>308	+	–	E	3	IVS3-3C>G (Hom)		Liver Ca,Tx
3	2	Sepsis, HM	8.7	Abnormal	23/56	5.4	71.028	+	NA	E	3	IVS3-3C>G (Hom)		Exitus
4	18	HM	4.4	Normal	36/57	0.3	550.6	+	+	L	3	IVS3-3C>G (Hom)		Well
5	45	Sibling hist.	7.5	NA	20/45	1.3	231.715	+	–	–[b]	3	IVS3-3C>G (Hom)		Well
6	2	Hepatomegaly	20	NA	36/114	3.19	78.273	+	NA	E	5	c.(440–441) del 8 nt		Non follow up
7	9	Ascites,HM	9.5	Abnormal	35/93	3.1	50.884	+	+	L	6	IVS6-1G>T (Hom)		Non follow up
8	4 day	Sibling hist.	12.8	Abnormal	29/103	20.1	152.652	+	NA	–[b]	6	IVS6-1G>T (Hom)		Non follow up
9	12	Cr. diarrhea HM,CMP	24	Abnormal	31/75	2.9	37.615	+	+	L	6	IVS6-1G>T (Hom)		Non follow up
10	4	HM	NA	Abnormal	16/22	0.7	NA	+	+	E	6	IVS6-1G>T (Hom)		Liver Ca
11	9 y	Rickets	6.8	Normal	27/37	0.63	37.25	+	+	E	6	IVS6-1G>T (Hom)		Well
12	36	HM	10.9	Abnormal	35/62	0.6	5135	+	+	L	6	IVS6-1G>T (Hom)		Well
13	3.5 y	Fanconi syn.	9.2	Abnormal	38/49	NA	6987	+	+	L	6	IVS6-1G>T/?		Non follow up
14	6	HSM	6.0	Abnormal	20/84	1.43	35.000	+	+	E	6	IVS6-1G>T (Hom)		Well
15	1.5	Sibling hist.	9.9	NA	23/76	3.18	5031	–	–	–[b]	6/8	IVS6-1G>T/c.698A>T	D233V (CH)	Well
16	26 day	Sibling hist.	9.4	Abnormal	23/145	3.67	369.643	+	–	–[b]	6	c.497T>G	V166G (Hom)	Well
17	1.5	Sibling hist.	15.6	NA	18/49	3.44	686.200	+	NA	–[b]	6	c.497T>G	V166G (Hom)	Well
18	19 m	Cr. diarrhea	9.8	Abnormal	28/80	0.59	34.479	+	NA	L	6	c.520C>T	R174X (Hom)	Nor. crisis
19	3	Rickets	17	Abnormal	27/77	NA	NA	+	+	E	8	c.696C>A	N232K (Hom)	Liver Ca-Tx-Ex
20	3	Diarrhea-HM	9.6	Abnormal	27/90	3.2	97.625	+	–	E	3	c.698A>T	D233V (Hom)	Well
21	3.5	Diarrhea-HM	10.6	Abnormal	30/84	0.64	761.5	+	–	E	8	c.698A>T	D233V (Hom)	Well
22	4	HM, rickets	10.3	NA	44/60	NA	>533	+	+	E	8	c.698A>T	D233V (Hom)	Liver Ca,Tx,Well
23	4	Hip.rickets	9	NA	14/13	0.8	NA	+	+	E	8	c.698A>T	D233V (Hom)	Liver Ca,Ex
24	20 day	Liver failure	6.3	Abnormal	88/312	6.5	52.830	+	+	VE	8	c.698A>T	D233V (Hom)	Well
25	7.5	HM	6.7	Abnormal	44/101	1.25	109.761	+	+	E	8/12	c.1107delG/c.698A>T	N344Tfsx/D233V (CH)	Non follow up
26	6	HM	8.6	Abnormal	30/53	2.3	36.570	+	–	E	9	c.709C>T	R237X (Hom)	Well
27	13	HM	9.5	Abnormal	98/245	5.08	1200	+	+	L	9	c.709C>T	R237X (Hom)	Exitus
28	4	Rickets	7.7	NA	27/64	0.36	18.7	+	+	E	9	c.776T>A	V259D (Hom)	Liver Ca,Tx,Well
29	12	HM	15.4	Normal	26/42	1.1	13.688	+	+	L	9	IVS9+2T>C (Hom)		Liver Ca,Tx,Well
30	NA	NA	NA	NA	NA	NA	NA	NA	NA	NA	9	IVS9+2T>C (Hom)		Non follow up
31	4	Sibling hist.	2.6	NA	32/87	1.5	>29.340	+	–	–[b]	12	IVS12+5G>A (Hom)		Well
32	5	Hepatomegaly	9.1	Abnormal	52/102	–	435	+	NA	E	12	IVS12+5G>A (Hom)		Well

Normal values of liver transaminase: ALT (6–50 IU/L for 0–5 days, 35–140 for 1–19 years), AST (5-45IU/ml for 0–5 days, 15–55 IU/ml for 1–19 years)

NA Not available, *HM* Hepatomegaly, *CMP* Restrictive cardiomyapathy, *GIS* Gastro intestinal system, Tx. Liver transplantation, *VE* Very early form, *E* Early form, *L* Late form, *Hom* Homozygous, *CH* Compound heterozygous

[a]It is accepted as 1 ng of AFP is approximately equal to 1 mIU (James et al. 1981; Wu et al. 1981; Blohm et al. 1998)

[b]Marker showed the patients with sibling history and diagnosed by screening

Table 2 Amplification of *FAH* gene fragments with long PCR using four primer sets

Amplicon	Exon	Amplicon size (bp)	Sequence of the primers
PCR1	1	5069	F-primer: TGCCAAACAGGTTGAATACAGAGGTTGTAACT R-primer: ACTCTTCTAGACCAAAGCACTTACCGCAGAC
PCR2	2,3,4,5	9672	F-primer: AAAGAGGGCAGGGAAACAAACTTTGACTAAG R-primer: TCATCACTGCCAAGGACACTCATATAGACACT
PCR3	6,7,8,9,10	11635	F-primer: TTAATCCAATAAAGGAACCAAGGTGGGTAA R-primer: ACCCTGGAACTGACAGTGAGACAGACAT
PCR4	11,12,13,14	12901	F-primer: CTGAGTGTCTTGCAGACAGACCCGAGAT R-primer: TAGAGAGCAGGGGTTTTAGCTGAGTCAACA

the time of diagnosis. Patient 15, who was screened for tyrosinemia type I at the age of 1.5 months because of a positive history (an affected older sibling), did not excrete SA in his urine (according to gas chromatography-mass spectrometry (GC-MS) results) despite the fact that plasma tyrosine and alpha-fetoprotein levels were elevated. The diagnosis of this patient was made via mutation analysis. During follow-up, GC-MS analysis detected SA in the urine of patient 14 at age 4 months. Haagen and Duran (1987) reported a case with tyrosinemia type I in which SA results were negative at 7 months of age. They proved that this finding was due to the low sensitivity of the test method used, which implies that there is a need for more sensitive tests.

Blood tyrosine levels of the patients in this study ranged from 4.4 to 25.1 mg dL^{-1}. A high-peak tyrosine level (25.1 mg dL^{-1}) was observed in one patient carrying the homozygous c.191delA mutation. AFP levels were higher than the normal range in all the patients at the time of diagnosis; however, based on the variation in the age of the patients at the time of diagnosis, AFP levels seemed to decrease considerably in terms of newborn period as time passes even if any treatment was given. A meaningful correlation was not observed between AFP level, and the severity of symptoms and disease type.

Altered coagulation parameters (PT and PTT) indicative of hepatic synthesis dysfunction were observed in all but except three of the patients in this study. In addition, two patients with a sibling history did not have distinctive clinical findings, but did have abnormal coagulation parameters at the time of diagnosis. Interestingly, although AST levels were elevated in 25 patients, the level of ALT (a specific enzyme for liver tissue) was significantly elevated in only five patients. Bilirubin levels were near the normal range at the time of diagnosis in all the patients. In summary, altered levels of tyrosine, AFP, coagulation parameters (PT/PTT), and AST were considered sensitive biomarkers for the diagnosis of tyrosinemia type I.

In total, 12 different disease-causing mutations were observed in 32 Turkish patients with tyrosinemia type I (Table 1): 4 missense mutations, 1 nonsense mutation, 4 splicing mutations, and 3 deletion-type mutations. Of the 12 mutations detected, 6 (V166G, R174X, D233V, R237X, IVS6-1G>A, and IVS12+5G>A) have been previously reported, the others (N232K, V259D, c.191delA, IVS3-3C>G, c.(440–441)del8 nt, and IVS9+2T>C) are designated as novel. Each novel mutation was screened in 50 control individuals and all were negative. In this study, IVS6-1G>A, D233V, and IVS3-3C>G were the most common mutations, comprising 25%, 17.1%, and 12.5% of the mutant alleles, respectively.

It is well known that the IVS6-1G>A mutation is the most frequently seen nucleotide change in the Mediterranean region, and accounts for approximately 60% of the deleterious alleles in the *FAH* gene (Bergman et al. 1998; Arranz et al. 2002). D233V is known to be specific to the Turkish population, as it has not been reported in other ethnic groups thus far, which is in agreement with the present results (Rootwelt et al. 1994, 1996). Surprisingly, IVS12 +5G>A, which is reportedly the most prevalent mutation worldwide (about 25% of alleles), was detected homozygously in only 2 patients in this study. It was suggested that D233V mutation causes *FAH* dysfunction by directly affecting the Ca^{2+} ligand at the active site, and that V166G mutation occurring near the active site may cause misfolding by directly affecting the active site geometry (Timm et al. 1999). Similarly, it could be predicted that N232K located in strand 8 of the β-roll at the active site may impair Ca^{2+} ligands, as does the D233V mutation. On the contrary, it could be suggested that another novel mutation-V259D-located in strand 9 of the β roll at the active site may cause dysfunction via misfolding, as does V166G.

Large cohorts are necessary to establish genotype–phenotype relationships for infrequent mutations; however, genotype–phenotype relationships were evaluated for the three most common mutations (IVS6-1G>A, D233V, and IVS3-3C>G) observed in this study. On the basis of age of diagnosis, presenting symptoms, tyrosine levels, and biochemical markers, a meaningful genotype–phenotype relationship for these three mutations was not observed. These findings are consistent with the consensus that appears in the literature.

Hepatocellular carcinoma developed in 7 of our patients. All of these patients were treated irregularly and inadequately with NTBC. In addition, there was not any information available concerning NTBC plasma levels or urinary succinylacetone excretion at the time these patients received NTBC treatment. Among these 7 patients, 5 who carried N232K, D233V, V259D, and IVS3-3C>G mutations developed liver cancer between 10 and 12 years of age. The patient carrying the IVS9+2T>C mutation developed liver cancer at 6 years of age. The patient with the IVS6-1G>T mutation was 24 years old when diagnosed with liver cancer. The AFP level in this patient was normal (4.16 IU mL^{-1}), and AFP was not immunohistochemically detected in liver tissue, whereas histological test results of liver tissue were compatible with neoplasm. On the contrary, noticeable increases in the AFP levels were not observed in these patients at the time liver cancer was diagnosed.

As a consequence, the present findings suggest that AFP has limited value for the early detection of liver cancer in tyrosinemia type I patients. Five patients in this study with the D233V mutation-3 of whom were previously reported by Rootwelt et al. (1994) developed liver carcinoma during follow-up at our clinic. Although 5 patients are not sufficient to conclude that the D233V mutation is a risk factor for liver cancer, patients carrying this nucleotide change should be followed up closely for liver neoplasm.

In conclusion, this study presents 6 novel mutations that were observed in tyrosinemia type I patients. The microarray resequencing method used to prescreen nucleotide changes was confirmed to be a rapid and cost-effective method for screening a large number of samples. Moreover, results of this study indicate that some mutations might be associated with a high risk of developing liver neoplasm.

Acknowledgement This study was supported by the State Planning Organization of Turkey (project No: DPT 2006K120640).

Synopsis

The article describes different type of FAH mutations and their clinical outcomes in tyrosinemia type 1 patients.

References

Arranz JA, Piñol F, Kozak L, Pérez-Cerdá C, Cormand B, Ugarte M, Riudor E (2002) Splicing mutations, mainly IVS6-1(G>T), account for 70% of fumarylacetoacetate hydrolase (FAH) gene alterations, including 7 novel mutations, in a survey of 29 tyrosinemia type I patients. Hum Mutat 20:180–188

Awata H, Endo F, Tanoue A, Kitano A, Nakano Y, Matsuda I (1994) Structural organization and analysis of the human fumarylacetoacetate hydrolase gene in tyrosinemia type I. Biochim Biophys Acta 1226:168–172

Bergman AJ, van den Berg IE, Brink W, Poll-The BT, Ploos van Amstel JK, Berger R (1998) Spectrum of mutations in the fumarylacetoacetate hydrolase gene of tyrosinemia type 1 patients in northwestern Europe and Mediterranean countries. Hum Mutat 12:19–26

Blohm ME, Vesterling-Hörner D, Calaminus G, Göbel U (1998) Alpha 1-fetoprotein (AFP) reference values in infants up to 2 years of age. Pediatr Hematol Oncol 15(2):135–142

Chakrapani A, Holme E (2006) Disorder of tyrosine metabolism. In: Fernandes J, Saudubray JM, van den Berghe G, Walter JH (eds) Inborn error metabolic diseases, 4th edn. Springer, Heidelberg, pp 233–242

Elpeleg ON, Shaag A, Holme E, Zughayar G, Ronen S, Fisher D, Hurvitz H (2002) Mutation analysis of the FAH gene in Israeli patients with tyrosinemia type I. Hum Mutat 19:80–81

Haagen AAM, Duran M (1987) Absence of increased succinylacetone in the urine of a child with hereditary tyrosinemia type I. Inherit Metab Dis 10(suppl 2):323–325

Heath SK, Gray RG, McKiernan P, Au KM, Walker E, Green A (2002) Mutation screening for tyrosinaemia type I. J Inherit Metab Dis 25:523–524

James TW, Linda B, Karen S (1981) Serum alpha fetoprotein (AFP) levels in normal infants. Pediatr Res 15:50–52

Mitchell GA, Grompe M, Lambert M, Tanguay RM (2001) Hypertyrosinemia. In: Scriver C, Beaudet AL, Sly WS, Valle D (eds) The metabolic and molecular bases of inherited diseases, 8th edn. Mc Graw-Hill, New York, pp 3733–3774

Rootwelt H, Berger R, Gray G, Kelly DA, Coşkun T, Kvittingen EA (1994) Novel splice, missense, and nonsense mutations in the fumarylacetoacetase gene causing tyrosinemia type 1. Am J Hum Genet 55:653–658

Rootwelt H, Høie K, Berger R, Kvittingen EA (1996) Fumarylacetoacetase mutations in tyrosinemia type I. Hum Mutat 7:239–243

St-Louis M, Tanguay RM (1997) Mutations in the fumarylacetoacetate hydrolase gene causing hereditary tyrosinemia type I: overview. Hum Mutat 9:291–299

Timm DE, Mueller HA, Bhanumoorthy P, Harp JM, Bunick GJ (1999) Crystal structure and mechanism of a carbon-carbon bond hydrolase. Structure 7:1023–1033

Van Spronsen FJ, Thomasse Y, Smit GP, Leonard JV, Clayton PT, Fidler V, Berger R, Heymans HS (1994) Hereditary tyrosinemia type I: a new clinical classification with difference in prognosis on dietary treatment. Hepatology 20:1187–1191

Van Spronsen FJ, Smit GP, Wijburg FA, Thomasse Y, Visser G, Heymans HS (1995) Tyrosinaemia type I: considerations of treatment strategy and experiences with risk assessment, diet and transplantation. J Inherit Metab Dis 18:111–114

Wu JT, Book L, Sudar K (1981) Serum alpha fetoprotein (AFP) levels in normal infants. Pediatr Res 15(1):50–52

JIMD Reports
DOI: 10.1007/8904_2011_11

Received: 6 July 2010; Revised: 12 December 2010;
Accepted: 15 December 2010; Published online: 25 June
Communicated by: Viktor Kožich

Generalized Arterial Calcification of Infancy: Fatal Clinical Course Associated with a Novel Mutation in *ENPP1*

Silvia Galletti, Yvonne Nitschke, Anna M. Malavolti, Giulia Aquilano, Giacomo Faldella,
Luigi Corvaglia, and Frank Rutsch

Abstract Generalized arterial calcification of infancy (GACI) is a rare condition characterized by arterial calcification within the internal elastic lamina associated with intimal proliferation, leading to stenosis of great and medium-sized vessels. This disease, caused by mutations in multiple exons of *ENPP1*, frequently results in death in infancy. Nowadays, the most promising therapeutic compounds for this rare disease are bisphosphonates. We describe a case of GACI associated with the novel mutation c.653A>T (p.D218V) in *ENPP1* on both alleles. The male infant was delivered prematurely and developed heart failure, severe hypertension, and diffuse calcifications of all arterial districts. He was treated with etidronate (18 mg/kg/day); however, the clinical condition did not improve, and a resolution of calcifications was not observed. The infant died within the 6th month of life of ischemic heart failure. We conclude that even if the diagnosis of GACI is established early and bisphosphonate treatment is started early, the prognosis can be very poor.

Keywords Bisphosphonates · Heart failure · Myocardial infarction · Inorganic pyrophosphate

S. Galletti, A.M. Malavolti, G. Aquilano, G. Faldella, and L. Corvaglia
Department of Gynecological, Obstetric and Pediatric Sciences,
University Hospital of Bologna, Bologna, Italy

Y. Nitschke
Department of General Pediatrics, Muenster University Children's Hospital, Muenster, Germany

F. Rutsch (✉)
Klinik und Poliklinik für Kinder-und Jugendmedizin,
Universitätsklinikum Münster, Albert-Schweitzer Campus 1, 48149
Münster, Germany
e-mail: rutschf@ukmuenster.de

Introduction

Generalized arterial calcification of infancy (GACI, MIM #208000) is a rare condition characterized by arterial calcification within the internal elastic lamina associated with intimal proliferation leading to stenosis of great and medium-sized arteries. The disease frequently results in death in infancy due to progressive ischemic heart failure. Although some cases of prenatal diagnosis have been reported, some patients are diagnosed postmortem at the earliest (Eronem et al. 2001). Presentation after the neonatal period is also unusual. Familial occurrence has suggested a genetic basis for this disease, which has been shown to be autosomal recessive (Rutsch et al. 2001). Although most patients die within the first 6 months of life, there have been rare reports of spontaneous resolution of the calcifications and cases of successful treatment with bisphosphonates (Sholler et al. 1984, Gleason et al. 1994, Ciana et al. 2006, Rutsch et al. 2008, Ramjan et al. 2009). We describe a case of GACI associated with a novel mutation in *ENPP* associated with preterm delivery and death early in life despite etidronate therapy.

Case Report

The patient was the second child of consanguineous Moroccan parents. The mother is a 40-year-old woman, who had seven previous pregnancies: a 16-year-old healthy boy, two miscarriages, two voluntary interruptions, and two stillborn infants delivered at 28 and 30 weeks of gestation. Amniocentesis of the current pregnancy revealed a normal 46, XY karyotype. In the 28th week of gestation, maternal hypertension was detected and fetal ultrasound outlined a severe pulmonary stenosis, tricuspid regurgitation (TR), cardiomegaly, polyhydramnios, and poor fetal movements. The mother underwent an urgent cesarean section because of acute fetal distress. The infant had a birth weight of

1,094 g, APGAR score was 4/6/10 at 1, 5 and 10 min, respectively, and the baby was intubated and ventilated. Physical examination showed no dysmorphic features, a grade 1 systolic murmur, a heart rate of 165 bpm and poor peripheral pulses. The liver was palpable 0.5 cm below the lower costal margin. The arterial blood pressure was normal (48/23 mmHg). Echocardiography, performed on the first day of life, revealed poor ventricular function (ejection fraction 40%), severe left ventricular hypokinesia and hypertrophy, brightness and hyperreflexia of the aortic wall from the arch to the abdominal tract with a low abdominal aorta pulsatility index. Inotropic support was immediately started with dobutamine and dopamine. Laboratory evaluation on the first day of life revealed elevated inflammatory markers (leukocytes 26,2100/μl [5,000–21,000/μl]); CrP 5.97 mg/dl [<0.5 mg/dl]; platelets 32,000/μl [250,000–450,000/μl]), acute kidney (creatinine 212.2 μmol/l [53–106 μmol/l]) and hepatic injury (aspartate aminotransferase, AST 397 U/l [15–60 U/l]; alanine aminotransferase, ALT 51 U/l [5–25 U/l]) with indirect hyperbilirubinemia (252 μmol/l [< 103 μmol/l]) that required exchange transfusion. A chest X-ray showed multiple periarticular calcifications of the left elbow (Fig. 1a). Cerebral ultrasound revealed marked cerebral parenchymal hyperechogenicity. Since the first hours of life the infant was treated with intravenous antibiotics, immunoglobulins, and cortisone for suspected sepsis. This was associated with a general clinical improvement and an increase of the cardiac ejection fraction, allowing for extubation on the 6th day of life. Treatment with phenobarbital and captopril was begun on the 10th day of life when the infant developed arterial hypertension (mean arterial pressure between 70–80 mmHg) and tonic-clonic seizures. The electroencephalogram (EEG) showed synchronous and asynchronous bursts, delta brush, and long interburst inactivity. During treatment the seizures resolved, while the hypertension persisted. Further echocardiography showed myocardial hyperechogenic foci on the apical portion of the left ventricle, hyperechogenicity, and brightness of both the pulmonary and the aortic valve (Fig. 2). Similar findings were also detected in the pulmonary, celiac, and renal arteries. Total body X-ray revealed more periarticular calcifications (left radiocarpal and both tarsal joints) and calcific spots in the left brachial and both iliac and femoral arteries (Fig. 1b). Total-body computed tomography (CT) confirmed widespread arterial calcifications and calcifications of the left lobe of the liver and the right kidney (Fig. 1c). At the age of 4 weeks, GACI was suspected

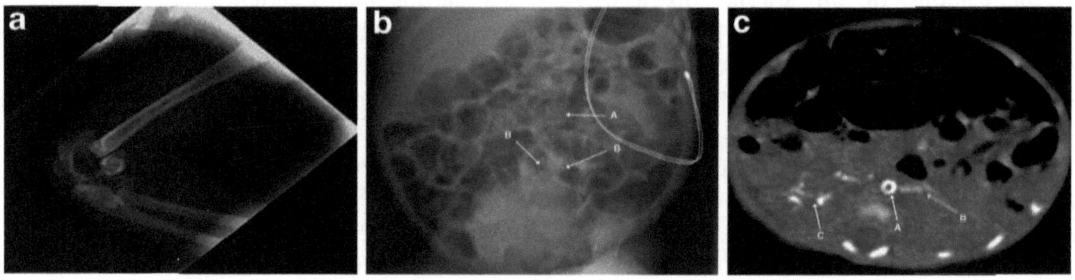

Fig. 1 *Radiographic manifestations of generalized arterial calcification of infancy. Panel A*: X-ray scan of the patient's left elbow. Note the calcification of the elbow joint. *Panel B*: Abdominal X-ray of the patient at 20 days of life. Note the calcification of the abdominal aorta (A) and of the bifurcation of the iliac arteries (B). *Panel C*: Chest CT of the patient at the age of 2 months. Note a ring-like aortic calcification (A) and spread calcifications over the left lobe of the liver (B) and in the right kidney (C)

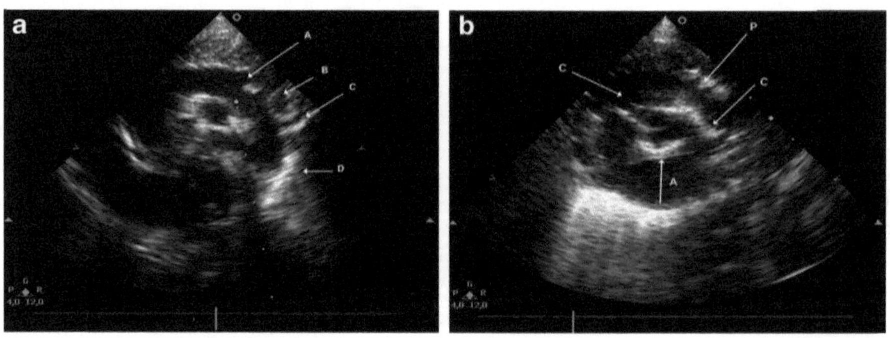

Fig. 2 *Echocardiographic manifestations of generalized arterial calcification of infancy. Panel A*: Echocardiogram at 1 month of life (upper-sternal axis view). Note the calcification of the aortic arch, descending aorta (D) and of the supraaortic vessels: brachiocephalic trunk (A), left carotid artery (B), left subclavian artery (C). *Panel B*: Echocardiogram at 1 month of life (parasternal short axis view). Note the hyperechogenicity of the pulmonary (P) and aortic (A) anulus and of the coronary arteries (C)

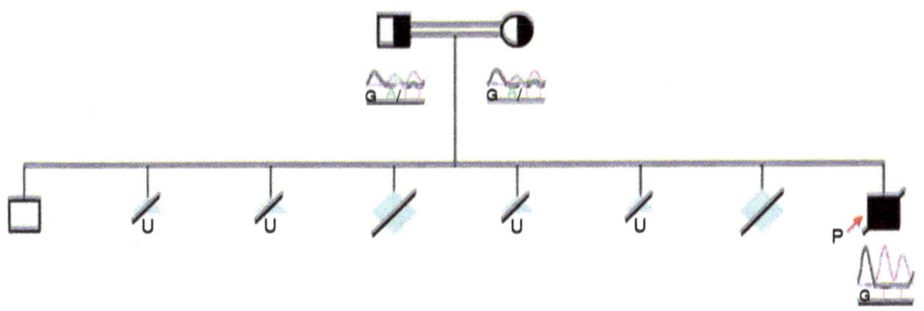

Fig. 3 *Pedigree tree of the family showing known genotypes and clinical phenotypes.* The affected allele is shaded *black*, whereas wild type alleles are *white*, unknown affected is shaded *gray*

after having excluded other potential causes of hyperechogenicity of the great vessels and hypertension. Because of congestive heart failure, which did not allow further fluid overload and because of limited venous access, an oral therapy with etidronate was started at the age of 1 month at a dose of 18 mg/kg body weight per day. Within the 2nd month of life, cerebral ultrasonography and Doppler studies revealed multicystic encephalomalacia with reduced arterial blood flow in the anterior and middle cerebral arteries bilaterally. At the age of 5 months, the infant was discharged on oral medication consisting of etidronate (18 mg/kg/day), phenobarbital (2.2 mg/kg × 2/day), captopril (0.3 mg/kg × 3/day), digoxine (0.016 mg/day), and vitamin D (900 U/day). Fifteen days later, the patient was readmitted for persistent vomiting and feeding refusal. On physical examination, the infant was tachypneic, tachycardic, and cyanotic. He started to have frequent episodes of desaturation with bradycardia requiring intubation. Echocardiography revealed a severely dilated cardiomyopathy and reduced ventricular function (EF less than 20%). Few hours later, he developed severe hypotension and died after a failed resuscitation attempt. Permission for autopsy was denied.

Genetic Analysis

With a set of 25 primer pairs, all 25 exons and their flanking splice sites of the *ENPP1* gene were amplified from genomic DNA by polymerase chain reaction (PCR). The PCR products were directly sequenced bidirectionally using an ABI 3730 DNA Analyzer and a BigDye Terminator v1.1 Cycle Sequencing Kit according to the manufacturer's protocol (Applied Biosystems). All primer sequences are available on request. Mutations were compared with the ENSEMBL polymorphism database. We detected the novel mutation c.653A>T (p.D218V) on exon 6 in *ENPP1* on both alleles in the patient's DNA. By analyzing the parental DNA, it was demonstrated that both parents carried the same mutation on one allele (Fig. 3).

Discussion

GACI is a rare autosomal recessive disorder due to systemic deficiency of nucleotide pyrophosphatase/phosphodiesterase 1 (NPP1) activity caused by mutations in multiple exons of *ENPP1* (MIM #173335), a gene located on chromosome 6q22–q23 (Rutsch et al. 2003). This gene encodes NPP1, a class II (intracellular NH_2 terminus) transmembrane glycoprotein ectoenzyme of 130 kDa with an extracellular domain containing two somatomedin B-like regions, a conserved calcium-binding EF hand, and a conserved phosphodiesterase/pyrophosphatase catalytic site (Terkeltaub 2001). The mutation c.653A>T (p.D218V) on exon 6 present on both alleles in our patient leads to a change of a conserved aspartate to a valine residue at position 218 in the catalytic side of NPP1.

The NPP1 cell surface enzyme regulates soft tissue calcification and joint cartilage mineralization by generating extracellular inorganic pyrophosphate (PP_i) (Rutsch et al. 2001). NPP1 is also involved both in the physiological inhibition of hydroxyapatite crystal growth and in chondrogenesis (Johnson et al. 2005). Deficiency of this enzyme therefore results in deposition of calcium hydroxyapatite crystals at the level of the lamina elastica interna of muscular arteries. Myointimal proliferation is found to be associated with calcified spots, but may also occur in areas lacking calcification and vice versa. This suggests a separate pathological process involved in the development of GACI (Dlamini et al. 2009). The aorta, renal, mesenteric, and carotid arteries are usually involved, while the cerebral vessels are commonly spared by the calcification process (Rutsch et al. 2008). Extravascular foci of periarticular calcification are often present. The first cases of GACI described were mostly recognized postmortem (Moran 1975). However, along with improving imaging technology in recent years, the diagnosis has frequently been established in early life and antenatal diagnosis has already been documented. Fetal ultrasound may reveal dilated cardiac ventricles, hydrops fetalis, and hyperechogenic large vessels (Levine et al. 2001). The most common clinical presentation, however, occurs within the neonatal period. Respiratory distress is the presenting feature in more than 50% of cases, followed by feeding intolerance, poor weight gain, tachypnea, tachycardia, lethargy, pallor, and cyanosis (Glatz et al. 2006, Rutsch et al. 2008). Patients may present with failure to thrive, hypertension, myocardial infarctions, and convulsions (Van der Sluis et al. 2006). Arterial biopsy is the gold standard for diagnosis, but this rather invasive technique can be replaced by combining standard imaging techniques (conventional radiographs, CT, and ultrasound). Recently, MRI and MR angiography (MRA) have been successfully used in the evaluation of patients with suspected GACI avoiding radiation exposure and potential renal toxicity associated with contrast enhanced CT examination (Pao et al. 1998; Tran and Boechat 2006). Widespread arterial calcifications can be related to numerous causes; thus, renal, osseous, parathyroid, and metabolic disorders should be ruled out before suspecting GACI (Patel et al. 2004). The differential diagnosis includes metastatic calcification secondary to advanced renal disease, arterial calcifications associated with anomalies of the heart and great vessels, hypervitaminosis D, primary or secondary hyperparathyroidism, Williams syndrome, syphilis, osteogenesis imperfecta, pseudoxanthoma elasticum, sepsis with myocarditis, endocarditis, coronary arteritis, rheumatic fever, polyarteritis nodosa, and also maternal conditions such as diabetes mellitus, allergy with asthma, and bronchiectasis (Glatz et al. 2006).

Few attempts to treat patients with GACI have been made with corticosteroids, estrogens, thyroid extract, and prostaglandins (Ciana et al. 1997). The most promising therapeutic compounds are bisphosphonates. Bisphosphonates are synthetic analogs of inorganic pyrophosphate, which block the conversion of calcium phosphate to hydroxyapatite and therefore may reduce ectopic calcification. So far, a standardized treatment approach for this rare disease could not be established. However, a recent multicenter genetic study and observational analysis has grouped 55 subjects affected by this disease. Seventeen patients have been treated with bisphosphonates, namely etidronate (10–20 mg/kg body weight per day p.o), pamidronate (0.1 mg/kg per week up to 5 mg/kg per day i.v.), clodronate or risedronate. Survival of these subjects was 65%, while 69% of patients not treated with bisphosphonates died during infancy (Rutsch et al. 2008). Treatment with aminobisphosphonates such as pamidronate disodium followed by oral risedronate was associated with a complete resolution of the calcifications and a good outcome (Ramjan et al. 2009). However, long-term survival has been reported also in patients receiving no specific anti-mineralisation therapy, suggesting that spontaneous resolution of calcification may also occur (Glatz et al. 2006). Etidronate, a non-nitrogen-containing compound among the bisphosphonates, has a stronger effect in inhibiting mineralization compared to the newer aminobisphosphonates and showed no adverse effect on growth (Van der Sluis et al. 2006). It is reasonable that the reduction of vascular calcifications increases vascular compliance and thereby reduces cardiac afterload. However, high-dose etidronate injections have been shown to induce vitamin D resistant rickets in rats (Atkin et al. 1988). Therefore, parameters of bone metabolism have to be closely monitored under etidronate therapy.

It has been hypothesized that hypophosphatemia may inhibit potential effects of defective NPP1-mediated PP_i generation, as PP_i and inorganic phosphate seem to have mutually antagonistic roles in tissue mineralization. In this respect, an interesting link has recently been reported (Lorenz-Depiereux et al. 2010, Levy-Litan et al. 2010) between inactivating mutations of *ENPP1* and hypophosphatemic rickets characterized by phosphaturia and elevated FGF23 levels. Moreover, previous studies on human subjects with *ENPP1* mutations have found a variable and incomplete reciprocal association between aberrant calcification and hypophosphatemic rickets (Ciana et al. 2006, Rutsch et al. 2008). Thus, inducing phosphaturia may have a potential benefit for GACI patients. This, however, has not become common practice. Nevertheless, bisphosphonates promote resolution of calcifications, but have no influence on the myointimal proliferation, which plays an important role in vascular stenoses (Rutsch et al. 2008).

In this particular case, the extreme prematurity of the infant, the severity and variability of clinical conditions since birth and the lack of diagnostic certainty may have delayed the diagnosis and thus the start of bisphosphonate therapy. Furthermore, during 4 months of therapy, a resolution of calcifications has not been observed. It may well be that the initiation of therapy might have been too late to resolve preexisting calcifications. Resistance to bisphosphonates may also be due to the particular mutation p.D218V in *ENPP1* affecting a conserved residue in the catalytic domain of the protein. Our experience again stresses the fact that GACI can present with a very rapid and fatal course. More data should be thus collected to better define prognosis and therapeutic possibilities.

Acknowledgment Y.N. and F.R are supported by a grant from the Interdisciplinary Center for Clinical Research, Münster University.

References

Atkin I, Ornoy A, Pita JC et al (1988) EHDP-induced rachitic syndrome in rats is not reversed by vitamin D metabolites. Anat Rec 220 (1):22–30

Ciana G, Colonna F, Forleo V et al (1997) Idiopathic arterial calcification in infancy: effectiveness of prostaglandin infusion for treatment of secondary hypertension refractory to conventional therapy: a case report. Pediatr Cardiol 18:67–71

Ciana G, Trappan A, Bembi B et al (2006) Generalized arterial calcification in infancy: two siblings with prolonged survival. Eur J Pediatr 165:258–263

Dlamini N, Splitt M, Durkan A et al (2009) Generalized arterial calcification of infancy: phenotypic spectrum among three siblings including one case without obvious arterial calcifications. Am J Med Genet A 149A:456–460

Eronem M, Pohjavuori M, Heikkila P (2001) Fatal outcome of two siblings with idiopathic arterial calcification of infancy diagnosed in utero. Pediatr Cardiol 22:167–169

Glatz AC, Pawel BR, Hsu DT et al (2006) Idiopathic Infantile arterial calcification: two case reports, a review of the literature and a role for cardiac transplantation. Pediatr Transplant 10:225–233

Gleason MM, Weber HS, Curan SE et al (1994) Idiopathic infantile arterial calcinosis: intermediate-term survival and cardiac sequelae. Am Heart J 127:691–695

Johnson K, Polewski M, van Etten D et al (2005) Chondrogenesis mediated by PPi depletion promotes spontaneous aortic calcification in NPP1-/- mice. Arterioscler Thromb Vasc Biol 25(4):686–691

Levine JC, Campbell J, Nadel A (2001) Prenatal diagnosis of idiopathic infantile arterial calcification. Circulation 103:325–326

Levy-Litan V, Hershkovitz E, Avizov L et al (2010) Autosomal-recessive hypophosphatemic rickets is associated with an inactivation mutation in the ENPP1 gene. Am J Hum Genet 86:273–278

Lorenz-Depiereux B, Schnabel D et al (2010) Loss-of-function ENPP1 mutations cause both generalized arterial calcification of infancy and autosomal-recessive hypophosphatemic rickets. Am J Hum Genet 86:267–272

Moran JJ (1975) Idiopathic arterial calcification of infancy: a clinicopathologic study. Pathol Annu 10:393–417

Pao DG, DeAngelis GA, Lovell MA (1998) Idiopathic arterial calcification in infancy: sonographic and magnetic resonance findings with pathologic correlation. Pediatr Radiol 28:256–259

Patel M, Savvas A, Rustum S et al (2004) Idiopathic arterial calcification in childhood. Pediatr Radiol 34:652–655

Ramjan KA, Roscioli T, Rutsch F et al (2009) Generalized arterial calcification of infancy: treatment with bisphosphonates. Nat Clin Pract Endocrinol Metab 5:167–172

Rutsch F, Vaingankar S, Johnson K et al (2001) PC-1 nucleoside triphosphate pyrophosphohydrolase deficiency in idiopathic infantile arterial calcification. Am J Pathol 158:543–554

Rutsch F, Ruf N, Vaingankar S et al (2003) Mutations in ENPP1 are associated with "idiopathic" infantile arterial calcification. Nat Genet 34(4):379–381

Rutsch F, Böyer P, Nitschke Y et al (2008) Hypophosphatemia, hyperphosphaturia, and bisphosphonate treatment are associated with survival beyond infancy in generalized arterial calcification of infancy. Circ Cardiovasc Genet 1:133–140

Sholler GF, Yu JS et al (1984) Generalized arterial calcification in infancy: three case reports, including spontaneous regression with long-term survival. J Pediatr 105:257–260

Terkeltaub RA (2001) Inorganic pyrophosphate generation and disposition in pathophysiology. Am J Physiol Cell Physiol 281: C1–C11

Tran KH, Boechat MI (2006) Idiopathic infantile arterial calcification: imaging evaluation and the usefulness of MR angiography. Pediatr Radiol 36:247–253

Van Der Sluis IM, Boot AM, Vernooij M et al (2006) Idiopathic infantile arterial calcification: clinical presentation, therapy and long-term follow-up. Eur J Pediatr 165:590–593

JIMD Reports
DOI: 10.1007/8904_2011_12

Received: 12 August 2010; Revised: 3 November 2010;
Accepted: 10 December 2010; Published online: 22 June
Communicated by: Ronald Wanders

Lymphoblastoid Cell Lines for Diagnosis of Peroxisome Biogenesis Disorders

Sabine Grønborg, Ralph Krätzner, Hendrik Rosewich, and Jutta Gärtner

Abstract Peroxisome biogenesis disorders (PBDs) are a group of autosomal-recessive developmental and progressive metabolic diseases leading to the Zellweger spectrum (ZS) phenotype in most instances. Diagnosis of clinically suspected cases can be difficult because of extensive genetic heterogeneity and large spectrum of disease severity. Furthermore, a second group of peroxisomal diseases caused by deficiencies of single peroxisomal enzymes can show an indistinguishable clinical phenotype. The diagnosis of these peroxisomal disorders relies on the clinical presentation, the biochemical parameters in plasma and erythrocyte membranes, and genetic testing as the final step. Analysis of patients' cells is frequently required during the diagnostic process, e.g., for complementation analysis to identify the affected gene before sequencing. In the cases with unclear clinical or biochemical presentation, patients' cells are analyzed to prove PBD or to demonstrate biochemical abnormalities that might be elusive in plasma. Cell lines from skin fibroblast that are usually generated for diagnostic workup are not available in all instances, mainly because the required skin biopsy is invasive and sometimes denied by parents. An alternative cellular system has not been analyzed sufficiently. In this study, we evaluated the alternative use of lymphoblastoid cell lines (LCLs), derived from a peripheral blood sample, in the diagnostic process for PBD. LCLs were suitable for immunofluorescence visualization of peroxisomal enzymes, complementation analysis, and the biochemical analysis to differentiate between control and PBD LCL. LCLs are therefore an easily obtainable alternative cellular system for a detailed PBD diagnostic workup with a reliability of diagnostic results equal to those of skin fibroblasts.

S. Grønborg (✉), R. Krätzner, H. Rosewich, and J. Gärtner
Department of Pediatrics and Pediatric Neurology, University Medical Center Göttingen, Georg August University, Robert-Koch-Strasse 40, 37075 Göttingen, Germany
e-mail: s.weller@med.uni-goettingen.de

Keywords Diagnosis · D-bifunctional protein deficiency · LCL · Peroxisome biogenesis disorder · Zellweger syndrome

Introduction

Due to their clinical and pathogenetic heterogeneity, inherited disorders of peroxisomal metabolism pose a diagnostic challenge. Peroxisomal diseases can be subdivided into two groups: PBD or group I peroxisomal diseases lead to a loss of function of most or all peroxisomal metabolic activities. Group II comprises diseases that are caused by deficiencies of single peroxisomal proteins only. More recently, a combined defect of peroxisomal and mitochondrial fission caused by mutation of a fission factor (DLP1) common to both organelles has added an inherited peroxisomal disease (Waterham et al. 2007). Different reports have emphasized mild and variant manifestations, especially of PBD, that are likely to elude diagnostic screening or even consideration as peroxisomal disease (Dursun et al. 2009; Raas-Rothschild et al. 2002; Zeharia et al. 2007).

Clinically, PBD patients present with Zellweger syndrome spectrum (ZSS) or, in 20% of cases, with rhizomelic chondrodysplasia punctata. Typical symptoms of ZSS include facial dysmorphism with high forehead, broad nasal bridge, external ear deformities, and large fontanelles, as well as severe neurological impairments such as muscular hypotonia, failure to thrive, seizures, and developmental deficiencies. Disease severity varies and life expectancy ranges from under 1 year in patients with classical Zellweger syndrome to survival into adulthood (Weller et al. 2003). Peroxisomal single protein deficiencies as of D-bifunctional protein (DBPD) or acyl-CoA oxidase show phenotypes resembling ZSS or have a distinct clinical picture as X-linked adrenoleukodystrophy (XALD), the most common peroxisomal disease (Wanders and Waterham 2006).

A common approach to the diagnosis of peroxisomal disorders is the measurement of peroxisomal metabolites, which accumulate or deplete in plasma or red blood cell membranes in case of a peroxisomal disease. Very long chain fatty acids (VLCFA) are measured as a screening tool for peroxisomal diseases in general, but may be normal in spite of the presence of peroxisomal disease (Ferdinandusse et al. 2006; Rosewich et al. 2006). A more detailed biochemical analysis and measurements in cultured patient cells, typically fibroblasts, can avoid this pitfall. Bile acid intermediates, phytanic and pristanic acid as well as plasmalogen measurements, help to differentiate between the two groups of peroxisomal diseases and direct suspicion to certain forms in the case of single protein deficiencies. Visualization of peroxisomes and the cellular distribution of peroxisomal matrix enzymes in patients' skin fibroblast cell lines is needed to establish the diagnosis of PBD and is included in a recent diagnostic flowchart for ZSS (Krause et al. 2009). It is also a prerequisite to diagnose the peroxisomal disorder with altered peroxisomal morphology due to a fission factor deficiency (Waterham et al. 2007) and possible further, related pathologies. The genetic heterogeneity of PBD – at least 13 different human genes are involved in peroxisome biogenesis and can cause the disease if mutant – makes complementation analysis in patient cell lines a rational tool to identify the underlying genetic defect (Krause et al. 2009).

In one patient referred to our center with suspected peroxisomal disease but inconclusive biochemical parameters in plasma (Grønborg et al. 2010), the patient's parents declined skin biopsy to establish a fibroblast cell line. With the parents' consent, we turned toward a lymphoblastoid cell line (LCL) that could be derived from a peripheral blood sample without any further burden to the patient. In one previous report, LCL from a PBD patient and a control individual were evaluated regarding cellular catalase distribution, VLCFA levels, and plasmalogen synthesis (Santos et al. 1993). To establish the usefulness of LCL for the diagnosis of peroxisomal disorders and PBD in particular, we studied immunofluorescence staining of peroxisomal membrane and matrix proteins, exemplified complementation analysis in PBD, and analyzed the biochemical profile of VLCFA, phytanic acid, and plasmalogens in LCL from three patients with peroxisomal disease and 19 control individuals. A clear-cut diagnosis could be confirmed for two patients with PBD relying on the biochemical profile, immunofluorescence, and complementation analysis in LCL.

Materials and Methods

Culture of LCLs

Lymphocytes from patients and controls were immortalized by standard Epstein–Barr virus transformation techniques using 5 ml of heparinized blood per patient (Neitzel 1986). LCLs were cultured in RPMI1640 medium supplemented with 25% FCS, 2 mM L-glutamine, 1% phythemagglutinin, 100 units/ml penicillin, and 100 μg/ml streptomycin at 37°C and 5% CO_2 following standard protocols.

Immunofluorescence Staining of LCL

Immunofluorescence staining was performed as described previously (Grønborg et al. 2010). Primary antibodies were rabbit anti-catalase (1:2,000), rabbit anti-PEX14 (1:1,000), and mouse anti-myc antibody (1:500) (Oxis International Inc., Foster City, CA, USA; Protein Tech Group, Chicago, IL, USA; BD Bioscience, Franklin Lakes, NJ, USA). Secondary antibodies were Alexa Fluor 488 goat anti-rabbit IgG (1:300), Texas Red goat anti-rabbit IgG (1:200), and Alexa Fluor 488 goat anti-mouse IgG (1:2,000) (Molecular Probes, Carlsbad, CA, USA). Subsequently, cells were washed and mounted with ProLong Gold Antifade Reagent with DAPI (Invitrogen GmbH, Karlsruhe, Germany) and analyzed under an Axio Imager M1 fluorescence microscope (Carl Zeiss MicroImaging GmbH, Göttingen, Germany).

Complementation Analysis

Transfections of LCL were conducted by electroporation using the Nucleofector Technology (Lonza, Basel, Switzerland); 1×10^6 LCLs were transfected with 2 μg total DNA per transfection, essentially following the manufacturers protocol. As a marker for peroxisomal protein import, the plasmid mycPECI for expression of peroxisomal Δ3, Δ2-enoyl-CoA isomerase (Geisbrecht et al. 1999) was transfected together with empty expression plasmid (pcDNA3.1; Invitrogen, Carlsbad, USA) or a plasmid for the expression of human PEX6 (pTY3) (Yahraus et al. 1996), respectively. PECI is a peroxisomal matrix protein with the conserved C-terminal peroxisomal import signal –SKL (–serine-lysine-leucine). Twenty-four hours after transfection, cells were processed for immunofluorescence as described above. The localization of mycPECI in the cytosol or in peroxisomes was determined by microscopy to prove complementation.

Biochemistry

For biochemical analysis, cells were pelleted, washed with medium, and resuspended at a concentration of 1×10^7 cells per 0.3 ml in NaCl 0.9%. For biochemical analysis, 100 μl of cell suspension was used. Each analysis was performed in replicate. Selected control and patients' samples were measured in two to three independently collected samples with two technical replicates each. For quantification of VLCFA and phytanic acid, we used the method of Hunnemann and

Hanefeld (1988). Briefly, VLCFA and phytanic acid were derivatized with acetylchloride as methylesters and calibrated via GC/MS against internal deuterated standards. Plasmalogens were determined accordingly, but 50 μl of cell suspension were used and levels were quantified relatively to C16:0 and C18:0 (Duran and Wanders 2008).

Patient and Control Cell Lines

Nineteen LCLs were used as control cell lines in total. Control cell lines were derived from healthy individuals and from patients with apparent nonmetabolic diseases. Patient LCLs were derived from two patients with peroxisomal biogenesis disorders (PBD1 and PBD2) and one patient with D-bifunctional protein deficiency (DBPD). The clinical phenotype and peroxisomal plasma biochemistry of the DBPD patient are described in a recent case report (Grønborg et al. 2010). This patient's peroxisomal parameters in plasma were only discretely conspicuous. PBD1 showed the symptoms of classical Zellweger syndrome and died at the age of 7 months. PBD2 has a milder phenotype and is currently 8 years old. Blood biochemistry from these patients pointed unambiguously toward a PBD with elevated VLCFA and reduced plasmalogen levels in erythroyte lipids (C22:0: PBD1 12.1, PBD2 14.4, normal 15–113 nmol/ml; C24:0: PBD1 27.5, PBD2 17.9, normal 12–94 nmol/ml; C26:0: PBD1 8.8, PBD2 3.45, normal 0.2–1.6 nmol/ml; C24:0/C22:0: PBD1 2.27, PBD2 1.25, normal 0.55–1.05; C26:0/C22:0: PBD1 0.727, PBD2 0.24, normal 0.005–0.029; phytanic acid: PBD1 1.7, PBD2 34.7, normal 0.3–9 nmol/ml; pristanic acid: PBD1 0.6, PBD2 19.9, normal 0–2 nmol/ml; C16:0 plasmalogens × 100/C16 fatty acids: PBD1 0.3, PBD2 4.4, normal 6.8–11.9; C18:0 plasmalogens × 100/C18 fatty acids: PBD1 0.2, PBD2 7.9, normal 10.6–24.9). Both PBD patients had previously been diagnosed on the molecular level by sequencing of the *PEX1* gene [PBD2: p.G843D/R872X; (Rosewich et al. 2005)] or by cDNA complementation analysis in fibroblasts and sequencing of the *PEX6* gene (PBD1: p.Asp865_Phe890del, unpublished) and were used in this study to examine the diagnostic possibilities in LCL.

Results

Immunofluorescence Staining of Peroxisomal Markers

Immunofluorescence staining of peroxisomal structures can visualize a protein import defect in patients with PBD and is frequently required to confirm or exclude the diagnosis (Krause et al. 2009). In the case of PBD, catalase or other peroxisomal matrix enzymes are not detected in punctate peroxisomal structures but evenly distributed in the cytosol. Peroxisomal membrane proteins (PMPs) can be found correctly localized or completely absent. Cells with group II peroxisomal disorders do not show mislocalization of peroxisomal proteins but might show fewer and enlarged peroxisomes (Huyghe et al. 2006). Peroxisomal morphology can be judged and a peroxisomal fission defect can be excluded. In LCL derived from a healthy control individual, a patient with D-bifunctional protein deficiency and a patient with PBD, immunofluorescence staining against PEX14, a PMP, revealed a punctate staining pattern in all three cell lines (Fig. 1, upper row). Staining against catalase, a peroxisomal matrix enzyme, clearly highlighted the patient with PBD in which anti-catalase staining pattern is distributed evenly throughout the cytosol in contrast to punctate staining in control and DBPD LCL (Fig. 1, lower row). No clear difference of peroxisome size and number could be detected comparing control LCL with LCL from the two patients.

Complementation of Peroxisome Biogenesis in a PBD Cell Line

Complementation analysis is suitable to determine which of the 12 known *PEX* genes accounts for ZSS and should be analyzed by direct sequencing (Krause et al. 2009). Figure 2 exemplifies complementation of peroxisomal protein import in an LCL derived from a patient with PBD: Cell line PBD1 was transfected with an expression construct for mycPECI (Geisbrecht et al. 1999), which is a peroxisomal matrix enzyme and serves as reporter for intact peroxisome biogenesis. Cotransfection of mycPECI with empty cDNA expression vector results in a cytosolic distribution of the protein within the cell (a); while cotransfection with PEX6 cDNA expression vector rescued the import defect, mycPECI is localized in punctate structures colocalizing with PEX14 (b), indicating complementation of the defective gene. Panels (c) and (d) show LCL from a control individual (d) and a patient with D-bifunctional protein deficiency (c) after transfection with mycPECI expression construct. Immunofluorescence analysis reveals colocalization of mycPECI and PEX14 as shown by the yellow color in the merged picture (right column).

Measurement of Peroxisomal Metabolites in LCL

Table 1 summarizes the concentrations of VLCFA (C22:0, C24:0, C26:0, C26:0/C22:0, C24:0/C22:0), phytanic acid, and plasmalogens in LCL of controls and patients with peroxisomal disorders. In 19 or 16 control individuals

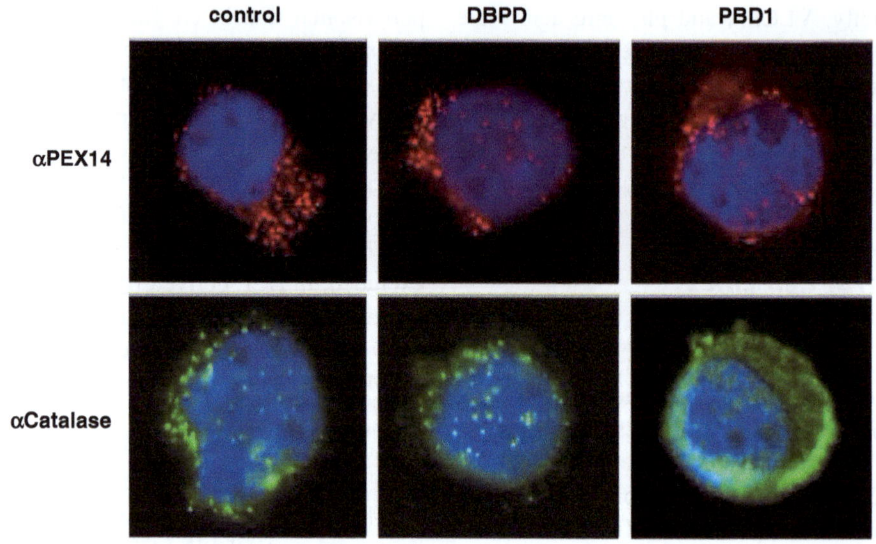

Fig. 1 Immunofluorescence staining of peroxisomal markers. LCL from a control person (*left*), a patient with D-bifunctional protein deficiency (DBPD, *middle*), and a patient with PBD (PBD1, *right*) were stained against the peroxisomal membrane protein PEX14 (*top*) and the peroxisomal matrix protein catalase (*bottom*) individually, using indirect immunofluorescence staining. PEX14 staining results in a punctate, peroxisomal staining pattern in all three cell lines, whereas catalase staining reveals a cytoplasmic distribution of the protein in the cell line with deficiency of peroxisome biogenesis only

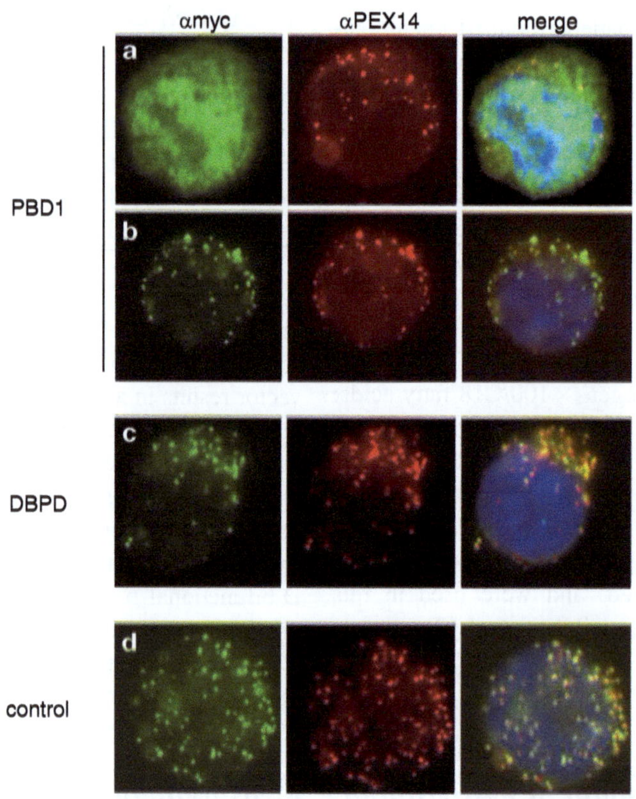

Fig. 2 Complementation of the peroxisome biogenesis defect. LCLs were transfected with a myc-tagged reporter construct for peroxisomal matrix protein import, mycPECI (Δ3, Δ2-enoyl-CoA-isomerase) (**a–d**) in addition to a vector expressing human PEX6 cDNA (**b**) or empty vector (**a**). Immunofluorescence staining visualizes the distribution of mycPECI in *green* (*left column*) and the peroxisomal membrane protein PEX14 in *red* (*middle column*). The *right column* shows the merged pictures to determine colocalization of the markers resulting in *yellow color*. Cells from a patient with PBD (PBD1) do not import mycPECI into peroxisomes resulting in a diffuse cytosolic distribution of the protein (**a**). After cotransfection of PBD1 cells with a vector expressing human PEX6 cDNA, mycPECI is imported into peroxisomes resulting in a punctate staining pattern of the protein and colocalization with PEX14 (**b**), confirming *PEX6* as the mutated gene in this cell line. Panels (**c**) and (**d**) show colocalization of peroxisomal matrix protein mycPECI and membrane protein PEX14 in LCLs from a patient with D-bifunctional protein deficiency and a control person

Table 1 Peroxisomal metabolites in lymphoblastoid cell lines

A

	C22	C24	C26	Phytanic acid	C24/C22	C26/C22
Controls (n=19)	7.19 (2.0/4.23–11.18)	10.84 (3.63/5.1–19.17)	1.89 (0.94/0.51–3.94)	1.21 (0.37/0.75–2.2)	1.49 (0.18/1.21–1.85)	0.26 (0.11/0.11–0.47)
DBPD	7.17 (0.39)	11.08 (0.41)	1.68 (0.09)	1.29 (0.21)	1.55 (0.03)	0.23 (0.01)
PBD1	6.63 (0.53)	13.12 (0.48)	7.94 (0.52)	1.61 (0.16)	1.98 (0.1)	1.21 (0.15)
PBD2	3.90 (0.25)	8.19 (1.11)	5.45 (1.01)	0.85 (0.06)	2.09 (0.16)	1.39 (0.18)

B

	C16 plasmalogens	C18 plasmalogens
Controls (n=16)	7.99 (2.23/4.15–11.35)	4.25 (1.1/2.25–5.6)
DBPD	6.38 (1.57)	3.0 (0.61)
PBD1	1.17 (0.12)	0.87 (0.15)
PBD2	1.88 (0.21)	1.13 (0.05)

Concentrations of C22:0, C24:0, and C26:0 fatty acids and phytanic acid in LCLs of controls (n=19) and three patients with DBPD and PBD (2×) (A). Values (nmol/1.67 × 10^7 cells) of controls are given as means with the standard deviation and the range of values in parentheses. Values of patients are given as means; the standard deviation is added in parentheses. Each control sample was measured in two to four replicates, selected control samples and patient PBD2 were measured in two to three independent biological samples, two replicates each.
C16 and C18 plasmalogen content in LCLs of controls (n=16) and three patients with DBPD and PBD (2×), respectively (B). The values are presented as described for A and are expressed as the ratio of C16 or 18 plasmalogens to C16 or 18 fatty acids, respectively, multiplied by 100. The values for each control sample were determined in replicate; selected control samples and all patient samples were measured in two to three independent biological samples, two replicates each.
C24/C22 and C26/C22 ratios clearly reveal the peroxisomal β-oxidation defect of VLCFAs in LCLs with peroxisome biogenesis disorders (PBD1 and PBD2) but not in a patient with DBP deficiency. Remarkably, this patient had only discretely conspicuous peroxisomal parameters in plasma (see text; Grønborg et al. 2010). Whether patients with DBPD and clear plasma biochemical phenotype will show pathological C24/C22 and C26/C22 ratios has yet to be shown. C16 and C18 plasmalogens are clearly reduced in patients with peroxisome biogenesis disorders (PBD1 and PBD2).

(for fatty acids or plasmalogens, respectively), these parameters were determined to define a normal range of values for the analyses in our laboratory (Table 1A, B). These can be contrasted to the results in LCL from one patient with DBPD and two patients with PBD (PBD1, PBD2). C22:0 values from patients DBPD and PBD1 lie well in the range of values for controls (4.23–11.18 nmol/ml). PBD2 has a slightly decreased value compared to controls (3.9 nmol/l). All three patients' cell lines have C24:0 values in the range of controls (5.1–19.17 nmol/l). C26:0 values from both patients with PBD (7.94 and 5.45 nmol/l, respectively) are outside the normal range (0.51–3.94 nmol/ml). The same is true for the values from C24/C22 and C26/C22 for PBD1 and PBD2. The deviation is more pronounced for C26/22 where the normal range of values is 0.11–0.47, while the PBD patients' values are 1.21 and 1.39, respectively. The DBPD patient does not have values outside the range determined for control individuals for any of the VLCFA analyses. It is important to note that the DBPD patient also had a mild clinical and plasma biochemical phenotype that led to diagnostic difficulties (Grønborg et al. 2010).

Phytanic acid measurement did not show abnormal results in both PBD patients. As phytanic acid is not an endogenous metabolite, its concentration in cells will depend on the phytanic acid content of the growth media. In fibroblasts systematically challenged with phytanic acid in the growth media, cells from PBD patients could be distinguished from control cells by indirectly measuring phytanic acid metabolization (Skjeldal et al. 1986).

C16 and C18 plasmalogen values from both PBD patients lie clearly outside the normal range, while the DBPD patient shows no deviation from the normal values, as expected (Table 1B).

Discussion

According to a recent diagnostic flowchart for the diagnosis of PBD, the evaluation of cultured cells becomes necessary when neither of the two most common mutations in the most frequently affected gene for PBD, namely G843D and c.2097-2098insT in *PEX1*, can be detected in the first diagnostic step (Krause et al. 2009). This applies only for those patients who present with typical biochemical changes in plasma (i.e., elevated VLCFA and decreased plasmalogens), otherwise immunofluorescence analysis in a fibroblast cell culture is required before the diagnostic flowchart is entered. Likewise, Steinberg et al. (2004) include patients into their diagnostic process only if the biochemical changes are clearly indicative of PBD, or if catalase immunofluorescence staining proved PBD and excluded a group II peroxisomal disease. Skin biopsy to obtain fibroblast cell culture might, although generally well accepted, be declined by parents because of its invasiveness or their beliefs as in one of our cases (Grønborg et al. 2010). In these cases and also in general, LCLs have the advantage to be easily obtained by transformation of blood lymphocytes sparing the patient a skin biopsy. Moreover, LCLs are immortal in contrast to primary fibroblast cell lines. A possible limitation is the required blood volume of 5 ml, especially in critically sick newborns and infants. LCL should thus represent an excellent alternative to fibroblasts in the diagnostic procedure for PBD that has so far not been studied in detail. We therefore evaluated the use of LCL in the diagnostic procedure for PBD.

Immunofluorescence staining clearly distinguished PBD patient LCL from LCL derived from a control individual and a patient with DBPD (Fig. 1), showing that clear-cut identification of PBD cells required for the initialization of different diagnostic procedures is possible in LCL with standard immunofluorescence staining. Previously, immunostaining of a peroxisomal matrix and membrane protein has been shown in LCL from one PBD patient and control (Santos et al. 1993). In addition to distinguishing between peroxisomal diseases of groups I and II, the quality and resolution of the immunostaining here should allow to demonstrate an altered peroxisomal morphology as present in patients with defects of peroxisomes and mitochondria caused by *DLP1* mutations (Waterham et al. 2007). As a screening tool, immunofluorescence analysis of peroxisomes in LCLs could be used to examine collections of LCL from patients with unclear dysmorphic and/or neurodegenerative syndromes with the potential to reveal further patients with abnormal peroxisomal morphology. Successful complementation analysis in LCL, exemplified for PBD1 and the complementation of the cellular phenotype with PEX6 cDNA (Fig. 2), makes LCLs a well-suited alternative starting material for the complete diagnostic program for PBD aiming at the identification of the genetic defect.

Plasma peroxisomal parameters can be normal in spite of the presence of peroxisomal disease, more frequently so in group II than in group I disorders (Ferdinandusse et al. 2006; Rosewich et al. 2006; Soorani-Lunsing et al. 2005; Wanders and Waterham 2006). Measurements of peroxisomal metabolites in cultivated cells, namely fibroblasts, have in some of these cases helped to elucidate the diagnosis of peroxisomal disease. Biochemical analysis of LCL in this report established that the profile of peroxisomal metabolites of PBD LCL can be diagnostic. C26:0, C24:0/C22:0, C26:0/C22:0, as well as C16 and C18 plasmalogens offered a clear diagnostic discrimination between values of controls and PBD patients 1 and 2 (Table 1). Whether LCL analysis of patients with ambiguous plasma values could lead to diagnosis has yet to be established. This was not the case for the DBPD patient in this study in whom plasma peroxisomal parameters had been discretely conspicuous only after repeated measurements (Grønborg et al. 2010). All values determined in LCL from this patient were within the normal range of values from control samples (Table 1). Further group II patients have to be studied to prove the reliability of peroxisomal metabolite measurements in LCL also for these patients. Ferdinandusse and colleagues describe DBPD patients with normal plasma C26:0 levels and C26/C22 ratio; however, in these cases elevated C26:0 levels were found in fibroblasts (Ferdinandusse et al. 2006).

Earlier studies of peroxisomal metabolites in cultivated cells from PBD patients included various reports in fibroblasts (Dacremont et al. 1995; Molzer 1993; Santos et al. 1993; Schutgens et al. 1993) and one report in LCL (Santos et al. 1993). For XALD patients, additional cell types have been analyzed, including blood leukocytes (Molzer 1993; Schutgens et al. 1993; Unterberger et al. 2007) and LCL in one report (Uto et al. 2008). Santos and colleagues observed C26/C22 levels in a control LCL (0.31) that were well comparable to the mean value of our study (0.26) (Santos et al. 1993). Increase of C26/C22 in the PBD LCL in their study (0.65) was less pronounced than for our patients (1.21/1.39), although the authors noted a generally higher amount of C26:0 accumulation in lymphoblast membranes than in fibroblasts. Indeed, C26/C22 values in control fibroblasts are approximately tenfold lower than in LCL in our and Santos' study (Santos et al. 1993) in some reports (Schutgens et al. 1993; Valianpour et al. 2003). Other analyses in fibroblasts have C26/C22 levels in fibroblasts comparable to our findings in LCL (Dacremont et al. 1995; Santos et al. 1993). Differences between these cell types might represent differences in membrane composition and/or culture conditions. Nevertheless, the deviations between studies emphasize the impact of different protocols for extraction and analysis of fatty acids.

In summary, study of LCL represents a reliable cellular system for the diagnostic process that is mandatory to establish the diagnosis of PBD. All diagnostic steps (immunofluorescence visualization of peroxisomes and peroxisomal enzymes, biochemical analysis of peroxisomal parameters, complementation analyses) that might be required to reach a definite, genetic diagnosis in this heterogeneous disease group can be performed in LCL and make these cells equivalent to fibroblasts in the diagnostic process. Thus, LCLs are a useful alternative to cultured skin fibroblasts, especially in cases where these cells are not available.

Acknowledgments We thank M. Schniewind and J. Kaiser for expert technical assistance, S. Thoms for help with microscopy, and R. Steinfeld, M. Henneke, and C. Brendel for sharing control cell lines. We thank C. Krause and A. Ohlenbusch for mutation analyses, D. Kube for access to the Nucleofector apparatus, and L. Florin (Bonn) for establishing LCL. This work has been supported by the Deutsche Forschungsgemeinschaft (Ga354/7-1) and the Bundesministerium für

Bildung und Forschung (BMBF) through the German Leukodystrophy Network.

Synopsis

Lymphoblastoid cell lines are on par with skin fibroblast cell lines for the diagnosis of peroxisome biogenesis disorders requiring immunofluorescence staining of peroxisomal matrix proteins, measurements of peroxisomal metabolites, and complementation analysis.

Details of Contributions of Individual Authors

Planning of the study: S.G., R.K., J.G.
Conduct of the study: S.G., R.K., H.R.
Reporting of work: S.G., R.K., J.G.

Name of Author Serving as Guarantor

S. Grønborg

Competing Interest

All authors declare that the answers to all questions on the JIMD competing interest form are No, and therefore they have nothing to declare.

Details of Funding

This work has been supported by the Deutsche Forschungsgemeinschaft (Ga354/7-1) and the Bundesministerium für Bildung und Forschung (BMBF) through the German Leukodystrophy Network. The authors confirm independence from the sponsors; the content of this chapter has not been influenced by the sponsors.

Details of Ethics Approval

Ethics approval was not required for this study.

References

Dacremont G, Cocquyt G, Vincent G (1995) Measurement of very long-chain fatty acids, phytanic and pristanic acid in plasma and cultured fibroblasts by gas chromatography. J Inherit Metab Dis 18(Suppl 1):76–83

Duran M, Wanders RJA (2008) Plasmalogens and polyunsaturated fatty acids. In: Blau N, Duran M, Gibson KM (eds) Laboratory guide to the methods in biochemical genetics. Springer, Berlin, pp 207–220

Dursun A, Gucer S, Ebberink MS, Yigit S, Wanders RJ, Waterham HR (2009) Zellweger syndrome with unusual findings: non-immune hydrops fetalis, dermal erythropoiesis and hypoplastic toe nails. J Inherit Metab Dis [Epub ahead of print]

Ferdinandusse S, Denis S, Mooyer PA et al (2006) Clinical and biochemical spectrum of D-bifunctional protein deficiency. Ann Neurol 59(1):92–104

Geisbrecht BV, Zhang D, Schulz H, Gould SJ (1999) Characterization of PECI, a novel monofunctional Delta(3), Delta(2)-enoyl-CoA isomerase of mammalian peroxisomes. J Biol Chem 274(31): 21797–21803

Grønborg S, Kratzner R, Spiegler J et al (2010) Typical cMRI pattern as diagnostic clue for D-bifunctional protein deficiency without apparent biochemical abnormalities in plasma. Am J Med Genet A 152A(11):2845–2849

Hunnemann DH, Hanefeld F (1988) Diagnose von peroxisomalen Erkrankungen – Erfahrungen mit einer empfindlichen massenfragmentographischen Bestimmung der sehr langkettigen Fettsäuren und der Phytansäure im Plasma. Monatsschr Kinderheilkd 136:529

Huyghe S, Mannaerts GP, Baes M, Van Veldhoven PP (2006) Peroxisomal multifunctional protein-2: the enzyme, the patients and the knockout mouse model. Biochim Biophys Acta 1761(9):973–994

Krause C, Rosewich H, Gartner J (2009) Rational diagnostic strategy for Zellweger syndrome spectrum patients. Eur J Hum Genet 17(6): 741–748

Molzer B (1993) Diagnosis of peroxisomal disorders with neurological involvement. Pädiatr Pädol 28(1):43–48

Neitzel H (1986) A routine method for the establishment of permanent growing lymphoblastoid cell lines. Hum Genet 73(4):320–326

Raas-Rothschild A, Wanders RJ, Mooijer PA et al (2002) A PEX6-defective peroxisomal biogenesis disorder with severe phenotype in an infant, versus mild phenotype resembling Usher syndrome in the affected parents. Am J Hum Genet 70(4):1062–1068

Rosewich H, Ohlenbusch A, Gartner J (2005) Genetic and clinical aspects of Zellweger spectrum patients with PEX1 mutations. J Med Genet 42(9):e58

Rosewich H, Waterham HR, Wanders RJ et al (2006) Pitfall in metabolic screening in a patient with fatal peroxisomal beta-oxidation defect. Neuropediatrics 37(2):95–98

Santos MJ, Moser AB, Drwinga H, Moser HW, Lazarow PB (1993) Analysis of peroxisomes in lymphoblasts: Zellweger syndrome and a patient with a deletion in chromosome 7. Pediatr Res 33(5): 441–444

Schutgens RB, Bouman IW, Nijenhuis AA, Wanders RJ, Frumau ME (1993) Profiles of very-long-chain fatty acids in plasma, fibroblasts, and blood cells in Zellweger syndrome, X-linked adrenoleukodystrophy, and rhizomelic chondrodysplasia punctata. Clin Chem 39(8):1632–1637

Skjeldal OH, Stokke O, Norseth J, Lie SO (1986) Phytanic acid oxidase activity in cultured skin fibroblasts. Diagnostic usefulness and limitations. Scand J Clin Lab Invest 46(3):283–287

Soorani-Lunsing RJ, van Spronsen FJ, Stolte-Dijkstra I et al (2005) Normal very-long-chain fatty acids in peroxisomal D-bifunctional protein deficiency: a diagnostic pitfall. J Inherit Metab Dis 28(6): 1172–1174

Steinberg S, Chen L, Wei L et al (2004) The PEX gene screen: molecular diagnosis of peroxisome biogenesis disorders in the Zellweger syndrome spectrum. Mol Genet Metab 83(3):252–263
Unterberger U, Regelsberger G, Sundt R, Bernheimer H, Voigtlander T (2007) Diagnosis of X-linked adrenoleukodystrophy in blood leukocytes. Clin Biochem 40(13–14):1037–1044
Uto T, Contreras MA, Gilg AG, Singh I (2008) Oxidative imbalance in nonstimulated X-adrenoleukodystrophy-derived lymphoblasts. Dev Neurosci 30(6):410–418
Valianpour F, Selhorst JJ, van Lint LE, van Gennip AH, Wanders RJ, Kemp S (2003) Analysis of very long-chain fatty acids using electrospray ionization mass spectrometry. Mol Genet Metab 79(3):189–196
Wanders RJ, Waterham HR (2006) Peroxisomal disorders: the single peroxisomal enzyme deficiencies. Biochim Biophys Acta 1763(12): 1707–1720
Waterham HR, Koster J, van Roermund CW, Mooyer PA, Wanders RJ, Leonard JV (2007) A lethal defect of mitochondrial and peroxisomal fission. N Engl J Med 356(17):1736–1741
Weller S, Gould SJ, Valle D (2003) Peroxisome biogenesis disorders. Annu Rev Genomics Hum Genet 4:165–211
Yahraus T, Braverman N, Dodt G et al (1996) The peroxisome biogenesis disorder group 4 gene, PXAAA1, encodes a cytoplasmic ATPase required for stability of the PTS1 receptor. EMBO J 15 (12):2914–2923
Zeharia A, Ebberink MS, Wanders RJ et al (2007) A novel PEX12 mutation identified as the cause of a peroxisomal biogenesis disorder with mild clinical phenotype, mild biochemical abnormalities in fibroblasts and a mosaic catalase immunofluorescence pattern, even at 40 degrees C. J Hum Genet 52(7):599–606

JIMD Reports
DOI: 10.1007/8904_2011_13

Received: 19 October 2010; Revised: 23 November 2010;
Accepted: 15 December 2010; Published online: 22 June
Communicated by: Jean-Marie Saudubray

First Report of a Molecular Prenatal Diagnosis in a Tunisian Family with Lysinuric Protein Intolerance

Nadia Esseghir, Chiraz Souissi Bouchlaka, Sondess Hadj Fredj, Amel Ben Chehida, Hatem Azzouz, Monique Fontaine, Neji Tebib, Marie Françoise Ben Dridi, Gilbert Briand, Taieb Messaoud, Amel Ben Ammar Elgaaied, and Naziha Kaabachi

Abstract Lysinuric protein intolerance (LPI, MIM# 222700) is an inherited aminoaciduria caused by defective transport of cationic amino acids (CAAs; arginine, lysine, ornithine) at the basolateral membrane of epithelial cells in the intestine and kidney. We report the first prenatal diagnosis by direct mutational analysis of LPI performed in a Tunisian family. An amniotic fluid sample was carried out at 16 weeks of gestation in a 32-year-old Tunisian woman who consulted for prenatal diagnosis. The 1471 delTTCT mutation at homozygous state was identified indicating that the fetus was affected by LPI. The identification of this specific mutation provides a tool, which can be easily applied in Tunisia for molecular diagnosis, genetic counseling, and prenatal diagnosis of LPI.

Keywords Direct mutational analysis · Lysinuric protein intolerance · Prenatal diagnosis

Lysinuric protein intolerance (LPI, MIM# 222700) is a rare autosomal recessive aminoaciduria caused by defective cationic amino acid (CAA; arginine, lysine, ornithine) transport at the basolateral membrane of epithelial cells in the intestine and kidney. Major symptoms include vomiting, diarrhea, failure to thrive, hepatosplenomegaly, bone marrow abnormalities, osteoporosis, episodes of coma, mental retardation, lung involvement (mainly as alveolar proteinosis), altered immune response, and chronic renal disease (Simell 2001). LPI has an autosomal recessive mode of inheritance and was first described in Finnish patients in 1965 (Perheentupa and Visakorpi 1965), with high prevalence of 1/60,000. It also occurs in Southern Italy and Japan. Sporadic cases have been described worldwide in 22 additional countries (Climbalistiene et al. 2007; Sperandeo et al. 2008). Five Tunisian patients (four families) were diagnosed with LPI, four of which have been previously described (Esseghir et al. 2006). LPI causes a severe clinical phenotype. Dietary regime and supplementation with citrulline may not be sufficient to avoid lethal complications. Prenatal diagnosis by linkage analysis of LPI is now available as reported by (Sperandeo et al. 1999). However, prenatal diagnosis by direct mutational analysis of LPI has not been previously reported in the literature. We report the first prenatal diagnosis by direct mutational analysis of LPI performed in a Tunisian Family.

A 32-year old woman (gravid 2, para 2) consulted for a prenatal diagnosis because her first child, an 11-year-old girl from consanguineous parents, is affected by LPI. The diagnosis of index case was made at the second year of life, on the occasion of Macrophage Activation Syndrome. Early signs of protein intolerance (vomiting, diarrhea, and selective protein aversion), poor growth and a pathological fracture at 18 months oriented the biochemical investigations revealing hyperammonemia and cationic amino aciduria with high urinary excretion of orotic acid, which confirmed the diagnosis of LPI (Table 1). Mutational analysis permitted the identification of the 1471 delTTCT mutation at homozygous state by DHPLC assay (Transgenomic) and was confirmed by direct sequencing (Applied Biosystem). For the prenatal diagnosis of the third pregnancy, an amniotic fluid sample was taken at 16 weeks of gestation. The 1471 delTTCT mutation at homozygous state was identified, indicating that the fetus was affected by LPI (Fig. 1).

N. Esseghir and N. Kaabachi (✉)
Biochemistry Laboratory, Rabta University Hospital, 1007 Jebbari, Tunis, Tunisia
e-mail: naziha.kaabachi@rns.tn; naziha.kaabachi@gmail.com

C.S. Bouchlaka and A.B.A. Elgaaied
Laboratory of Genetic, Immunology and Human Pathologies, EL Manar University, Tunis, Tunisia

S.H. Fredj and T. Messaoud
Laboratory of Biochemistry and Molecular Biology, Children Hospital, Tunis, Tunisia

A.B. Chehida, H. Azzouz, N. Tebib, and M.F.B. Dridi
Department of Pediatrics, Rabta University Hospital, Tunis, Tunisia

M. Fontaine and G. Briand
Biochemistry and Molecular Biology Laboratory, Research University Hospital Complex, Lille, France

Table 1 Biochemical data of the affected LPI girl

Parameters	Concentrations	
Plasma amonemia	73 μmol/L (10–60)	
Urinary orotic acid	103.5 mmol/mol creatinine (0.35–3.5)	
Amino acids	*Urine* (μmol/mmol of creatinine)	*Plasma* (μmol/L)
Ornithine	3.058 (0–80)	33 (29–105)
Lysine	2.517 (0–850)	41 (119–203)
Arginine	1.548 (0–70)	28 (28–101)

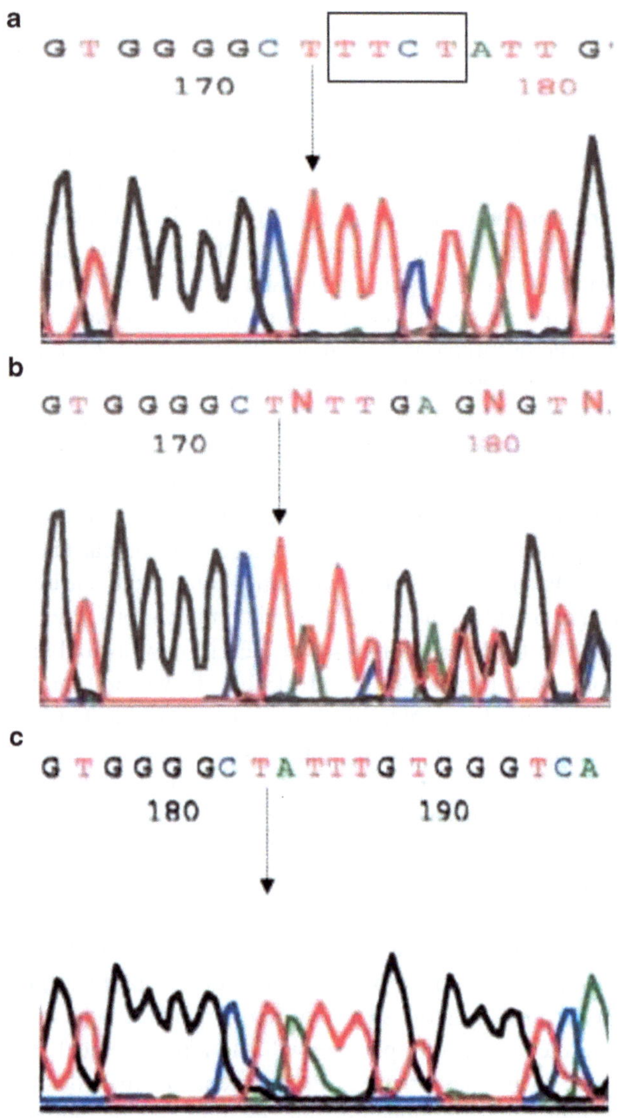

Fig. 1 The sequence electropherogram of the sense strand in exon 9: an *arrow* indicates the position's deletion. (**a**) The normal sequence from a control. (**b**) The heterozygous sequence from a patient's parent. (**c**) The homozygote delTTCT present in patient

The 1471 delTTCT mutation is one of 51 different SLCA7A-specific mutations that have been identified in 142 patients with LPI (Font-Llitjos et al. 2009). Subsequently, the same mutation was identified at homozygous state in five Tunisian patients with LPI (Esseghir et al. 2009). Previously, it had been found at homozygous state in two others Tunisian siblings (Sperandeo et al. 2000).

In conclusion, the 1471 delTTCT mutation, which has been identified in a total of seven Tunisian unrelated patients with LPI, seems to be the common mutation for this disorder in Tunisia. Considering these results, Tunisian patients suspected of LPI can first be screened for this specific mutation, allowing a cost-effective and rapid diagnosis of LPI in Tunisia.

References

Climbalistiene L, Lehnert W, Huoponen K, Kucinskas V (2007) First reported case of lysinuric protein intolerance (LPI) in Lithuania, confirmed biochemically and by DNA analysis. J Appl Genet 48: 277–280

Esseghir N, Tebib N, Bouchlakka Souissi C et al (2006) Lysinuric protein intolerance: clinical and biochimical profile of 4 Tunisian children. J Inherit Metab Dis 29:109

Esseghir N, Bouchlakka Souissi C, Tebib N et al (2009) The common c.1185-1188delTTCT mutation of SLC7A7 gene from Tunisian patients with lysinuric protein intolerance. Mol Genet Metab 98:145

Font-Llitjos M, Rodriguez-Santiago B, Espino M et al (2009) Novel SLC7A7 large rearrangements in lysinuric protein intolerance patients involving the same AluY repeat. Eur J Hum Genet 17: 71–79

Perheentupa J, Visakorpi JK (1965) Protein intolerance with deficient transport of basic amino acids: another inborn error of metabolism. Lancet 2:813–816

Simell O (2001) Lysinuric protein intolerance and other cationic aminoacidurias. In: Scriver CR, Beaudet AL, Sly WS, Valle D (eds) The metabolic and molecular bases of inherited disease. McGraw-Hill, New York, pp 4033–4956

Sperandeo MP, Buoninconti A, Passariello A et al (1999) Feasibility of prenatal diagnosis of lysinuric protein intolerance by linkage analysis: a case report. Prenat Diagn 19:771–773

Sperandeo M, Bassi MT, Riboni M et al (2000) Structure of the SLC7A7 gene and mutational analysis of patients affected by lysinuric protein intolerance. Am J Hum Genet 66:92–99

Sperandeo MP, Andria G, Sebastio G (2008) Lysinuric protein intolerance: update and extended mutation analysis of the SLC7A7 gene. Hum Mutat 29:14–21

JIMD Reports
DOI: 10.1007/8904_2011_14

Received: 6 September 2010; Revised: 5 November 2010;
Accepted: 30 December 2010; Published online: 22 June
Communicated by: Ed Wraith

Foot Process Effacement with Normal Urinalysis in Classic Fabry Disease

Takahiro Kanai, Takanori Yamagata, Takane Ito, Jun Odaka, Takashi Saito, Jun Aoyagi, Masahisa Kobayashi, Toya Ohashi, Yoshihiko Ueda, and Mariko Y Momoi

Abstract Fabry disease is an X-linked glycosphingolipidosis caused by deficient synthesis of the enzyme α-galactosidase A, which results in glycosphingolipidosis, predominantly globotriaosylceramide, progressively accumulating in systemic tissue. A dominant complication of Fabry disease is nephropathy. The average age for the development of clinical nephropathy is 27 years in male patients, with up to half of all patients developing end-stage renal failure by their 50s. A recent study revealed podocytes play important roles in antiproteinuria. Podocyte injury leads to foot process effacement and proteinuria. The foot process effacement induces podocyte depletion from the glomerular wall, glomerulosclerosis, and results in end-stage renal failure. We report on a 13-year-old boy with classic Fabry disease, who developed foot process effacement *and podocyte depletion* even before proteinuria appeared. At the time, his only symptom of Fabry disease was acroparesthesia. He was administered Agalsidase β (1 mg/kg/dose div) every other week and *14 months after treatment*, his renal function *remained normal*. This is the first report of a patient with classic Fabry disease, with only acroparesthesia, who had normal urinalysis but manifested foot process effacement and podocyte depletion. Podocytes are highly differentiated cells with a limited capacity for cell division and replacement. The large individual variation and often progressive nature of this disease raises concerns about the appropriate timing for initiating enzyme replacement therapy (ERT). Recent data have shown a limited effect of ERT on progressive organ damage. In our case, ERT was initiated before proteinuria appeared, with good outcome.

Keywords Enzyme replacement therapy · Fabry disease · Foot process effacement · Proteinuria · Renal pathology

Introduction

Fabry disease (MIM ID #301500) is an X-linked glycosphingolipidosis caused by deficient synthesis of the enzyme α-galactosidase A (EC 3.2.1.22). Glycosphingolipids, predominantly globotriaosylceramide (GL-3), progressively accumulate in systemic tissue (Eng et al. 2006). Nephropathy is a dominant complication of Fabry disease, and the average age for the development of clinical nephropathy is 27 years in male patients, with up to half of all patients developing end-stage renal failure by their 50s (Branton et al. 2002). GL-3 accumulation begins during fetal development, and cellular inclusions of GL-3 have been detected in fetal kidney podocytes (Desnick 2007). However, it remains unclear when cellular inclusion of GL-3 triggers foot process effacement from podocyte injury, causing *podocyte depletion from the glomerular wall* and end-stage renal failure (Wiggins 2007). We present a case of classic Fabry disease with foot process effacement *and podocyte depletion*, but with normal urinalysis results and acroparesthesia the only symptom of the disease.

T. Kanai (✉), T. Yamagata, T. Ito, J. Odaka, T. Saito, J. Aoyagi, and M.Y. Momoi
Department of Pediatrics, Jichi Medical University, 3311-1 Yakushiji, Shimotsuke-shi, Tochigi 329-0498, Japan
e-mail: tkanai@jichi.ac.jp

M. Kobayashi
Department of Pediatrics, The Jikei University School of Medicine, Tokyo, Japan

T. Ohashi
Department of Pediatrics, The Jikei University School of Medicine, Tokyo, Japan
and
Department of Gene Therapy, Institute of DNA Medicine, The Jikei University School of Medicine, Tokyo, Japan

Y. Ueda
Department of Pathology, Koshigaya Hospital, Dokkyo Medical University, Saitama, Japan

Case Report

A 13-year-old boy consulted at our hospital for fever- or exercise-associated burning pain on the soles and palms that had persisted for 3 years. His maternal grandfather, who had not received a diagnosis of Fabry disease, died of multi-organ failure when he was 45 years old. His mother and maternal grandfather had acroparesthesia when they were young. His father and maternal grandmother had not experienced any signs or symptoms of Fabry disease.

Physical Examination and Laboratory Data

The patient's height was 159.7 cm; body weight, 47.6 kg; blood pressure, 106/66 mmHg; and body temperature, 36.8°C. He had no rash or angiokeratomas. It was determined that he had a point mutation of C382Y on exon 7 of the α-galactosidase A gene. Examination of blood revealed extremely low α-galactosidase A activity [0.1 nmol/mg protein/h; reference level, 50–110 nmol/mg protein/h, *measured using a fluorogenic substrate* (*Mayes et al.* 1981)], elevated GL-3 level [17.5 μg/mL; reference level, less than 7.0 μg/mL, *measured by liquid chromatography/tandem mass spectrometry* (*Roddy et al.* 2005)], normal creatinine level (0.46 mg/dL) and blood urea nitrogen level (11 mg/dL), and no acidemia (pH 7.346; HCO^{3-}, 23 mEq/L). Repeated urinalysis showed negative proteinuria [less than 4 mg/dL and 50–80 mg/day, *measured by a pyrogallol red-molybdate complex* (*Watanabe et al.* 1986)], no hematuria (less than 5 per high-power field), and no hyposthenuria (specific gravity, 1.025–1.030 on the morning samples), but a slightly elevated β_2-microglobulin level (544.0 μg/L). *His glomerular filtration rate was 144 mL/min/1.73 m^2*. He presented with no other signs or symptoms of Fabry disease (e.g., hypohidrosis, gastrointestinal disturbances, thermal sensation loss, or cornea verticillata by slit-light examination). Echocardiography demonstrated normal function and structure. Cerebral involvement could not be detected by magnetic resonance imaging.

Renal Pathologic Findings with Normal Urinalysis

After receiving informed consent from the patient and his parents and approval from the Ethics Committee of Jichi Medical University, we performed a renal biopsy for renal damage assessment. Twenty-six glomeruli were observed, demonstrating diffuse global vacuolated podocytes by light microscopy, and diffuse global whorled inclusions in podocytes and focal segmental foot process effacement by electron microscopy (Fig. 1). The podocytes with whorled inclusions had foot processes and foot process effacement was detected where there were no podocytes. We observed no vacuolated endothelial cells or mesangial cells. No deposits were observed based on immunofluorescent studies (IgA, IgG, IgM, C3, C4, and C1q), and there was no glomerular sclerosis. Tubular cells were also vacuolated by light microscopy and included whorled inclusions diffusely and globally. Although there was no tubulo-interstitial fibrosis, the tunica media of the small arteries were vacuolated.

Agalsidase β (1 mg/kg/dose div) was administered every other week, which decreased the GL-3 level and provided relief from the acroparesthesia. *Fourteen months after treatment, the patient maintained a normal GL-3 plasma level (less than 7.0 μg/mL), normal renal function, normal urinalysis, and was free of other signs and symptoms of Fabry disease.*

Discussion

Foot process effacement *and podocyte depletion* had already developed in our classic Fabry disease case, in a patient who only presented with acroparesthesia, a first symptom of Fabry disease (Desnick and Brady 2004). This is the first report of a patient with classic Fabry disease, with only acroparesthesia, who had normal urinalysis but manifested foot process effacement from podocyte injury. Previous studies reported several patients with detectable proteinuria and renal damage such as foot process effacement and glomerular sclerosis (Gubler et al. 1978; Tondel et al. 2008).

The findings of diffuse global whorled inclusions in podocytes, focal segmental foot process effacement, podocytes with both whorled inclusions and foot processes, and foot process effacement without podocytes indicate that depletion of podocytes with whorled inclusions was induced by foot process effacement. Foot process effacement depends on disruption of the actin cytoskeletal network, which forms the slit membrane in the podocytes (Asanuma et al. 2007). This disruption results in proteinuria (Kawachi et al. 2009) and is thought to cause podocyte depletion (Wiggins 2007). Podocytes are highly differentiated cells with a limited capacity for cell division and replacement (Wiggins 2009), and podocyte injury and loss from the glomerulus result in end-stage renal failure by glomerular sclerosis (Wiggins 2009). In our case, progression of podocyte injury would have caused proteinuria, severe podocyte depletion, and ultimately end-stage renal failure.

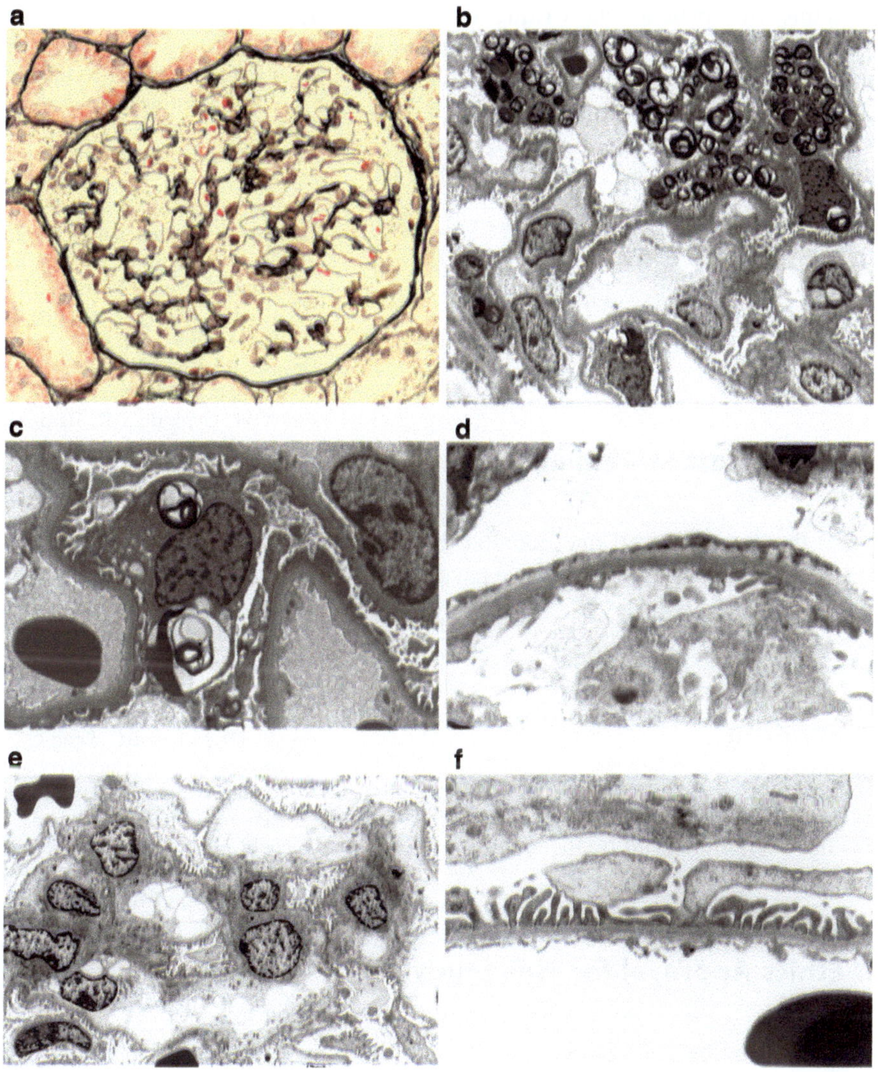

Fig. 1 (**a**) Vacuolated podocytes (original magnification ×400; periodic-methenamine silver stain). (**b**) Diffuse global whorled inclusions in podocytes and focal segmental foot process effacement (original magnification ×1,200). The podocytes with whorled inclusions had foot processes and foot process effacement was detected where there were no podocytes. (**c**) Whorled inclusions (*zebra body*) in podocytes with foot processes (original magnification ×6,000). (**d**) Foot process effacement that we observed where there were no podocytes (original magnification ×10,000). (**e**) *and* (**f**) *Normal foot process at the remission phase of minimal change nephrotic syndrome (original magnification ×2,500 and ×10,000) in a 13-year-old female*

The large individual variation and often progressive nature of this disease raise concerns about the appropriate timing for initiation of enzyme replacement therapy (ERT). Recent data have shown a limited effect of ERT on progressive organ damage when treatment is initiated at a later stage (Breunig et al. 2006). The findings in our case are informative for establishing the optimal ERT initiation time. Our patient showed no proteinuria, but already had foot process effacement resulting from podocyte injury *and podocyte depletion* that would lead to glomerular sclerosis. Therefore, in our case, ERT was appropriately initiated before proteinuria appeared, with good outcome.

Conclusions

Foot process effacement resulting from podocyte injury *and podocyte depletion from the glomerular wall* had already manifested when acroparesthesia developed in this patient. Although urinalysis was normal, it was a case of classic Fabry disease. Therefore, normal urinalysis does not necessarily indicate normal renal tissue. Accumulated data are needed for a better understanding of when to initiate ERT.

Synopsis

Normal urinalysis does not always indicate normal kidney tissue in Fabry disease.

Details of the Contributions of Individual Authors

Dr. Kanai, the author of this article and the doctor of this patient; Prof Yamagata and Prof Momoi, advisors for this article; Dr. Ito, Dr. Odaka, Dr. Saito, and Dr. Aoyagi, assisted in treating this patient and writing this article; Dr. Kobayashi and Prof Ohashi, detected the gene abnormality; and Prof Ueda, reported the pathologic findings.

Name of One Author Who Serves as Guarantor for This Article

Dr. Takahiro Kanai.

A Competing Interest Statement

We declare that the answers to all questions on the JIMD competing interest form are NO; therefore, we have nothing to declare.

Details of Funding

We confirm independence from the sponsors; the content of the article has not been influenced by the sponsors.

Details of Ethics Approval for This Study

This study was approved by The Ethics Committee of Jichi Medical University with the patient consent statement.

References

Asanuma K, Yanagida-Asanuma E, Takagi M, Kodama F, Tomino Y (2007) The role of podocytes in proteinuria. Nephrology (Carlton) 12(Suppl 3):S15–S20

Branton M, Schiffmann R, Kopp JB (2002) Natural history and treatment of renal involvement in Fabry disease. J Am Soc Nephrol 13 (Suppl 2):S139–S143

Breunig F, Weidemann F, Strotmann J, Knoll A, Wanner C (2006) Clinical benefit of enzyme replacement therapy in Fabry disease. Kidney Int 69(7):1216–1221

Desnick RJ (2007) Prenatal diagnosis of Fabry disease. Prenat Diagn 27 (8):693–694

Desnick RJ, Brady RO (2004) Fabry disease in childhood. J Pediatr 144 (5 Suppl):S20–S26

Eng CM, Germain DP, Banikazemi M et al (2006) Fabry disease: guidelines for the evaluation and management of multi-organ system involvement. Genet Med 8(9):539–548

Gubler MC, Lenoir G, Grunfeld JP, Ulmann A, Droz D, Habib R (1978) Early renal changes in hemizygous and heterozygous patients with Fabry's disease. Kidney Int 13(3):223–235

Kawachi H, Han GD, Miyauchi N, Hashimoto T, Suzuki K, Shimizu F (2009) Therapeutic targets in the podocyte: findings in anti-slit diaphragm antibody-induced nephropathy. J Nephrol 22(4):450–456

Mayes JS, Scheere JB, Sifers RN, Donaldson ML (1981) Differential assay for lysosomal alpha-galactosidases in human tissues and its application to Fabry's disease. Clin Chim Acta 112(2):247–251

Roddy TP, Nelson BC, Sung CC et al (2005) Liquid chromatography-tandem mass spectrometry quantification of globotriaosylceramide in plasma for long-term monitoring of Fabry patients treated with enzyme replacement therapy. Clin Chem 51:237–240

Tondel C, Bostad L, Hirth A, Svarstad E (2008) Renal biopsy findings in children and adolescents with Fabry disease and minimal albuminuria. Am J Kidney Dis 51(5):767–776

Watanabe N, Kamei S, Ohkubo A et al (1986) Urinary protein as measured with a pyrogallol red-molybdate complex, manually and in a Hitach 726 automated analyzer. Clin Chem 22:1151–1154

Wiggins RC (2007) The spectrum of podocytopathies: a unifying view of glomerular diseases. Kidney Int 71(12):1205–1214

Wiggins J (2009) Podocytes and glomerular function with aging. Semin Nephrol 29(6):587–593

JIMD Reports
DOI: 10.1007/8904_2011_15

Received: 14 September 2010; Revised: 8 November 2010;
Accepted: 15 December 2010; Published online: 22 June
Communicated by: Jean-Marie Saudubray

Growth Hormone Therapy Is Safe and Effective in Patients with Lysinuric Protein Intolerance

Harri Niinikoski, Risto Lapatto, Matti Nuutinen, Laura Tanner, Olli Simell, and Kirsti Näntö-Salonen

Abstract *Background*: Lysinuric protein intolerance (LPI) is an autosomal recessive cationic amino acid transport defect characterized by episodes of postprandial hyperammonemias and spontaneous protein aversion. Subnormal growth is common in spite of appropriate nutritional therapy. Growth hormone (GH) therapy promotes appetite, protein synthesis and accretion, but its possible growth-promoting effects and safety in patients with LPI are poorly known.

Methods: Four LPI children aged 7–16 years were treated with GH for a period of 3–4.5 years. Dietary intakes and plasma amino acid levels were analyzed frequently in addition to routine monitoring of GH therapy.

Results: Insulin-like growth factor-1 concentration was low and bone age was delayed in all LPI patients, but GH provocative test was pathological in only one of the patients. During the 3–4.5 years of GH therapy (dose 0.035–0.050 mg/kg/day), bone age did not catch up but height standard deviation score (SDS) improved by 0.7–1.8 SDS. There were no episodes of hyperammonemias.

Conclusions: Our data support safety and growth-promoting potential of long-term GH therapy in patients with LPI.

Keywords Children · Growth · Growth hormone · Lysinuric protein intolerance · Nutrition

H. Niinikoski (✉), L. Tanner, O. Simell, and K. Näntö-Salonen
Department of Pediatrics, University of Turku, Kiinamyllynkatu 4-8, Turku, Finland
e-mail: harri.niinikoski@tyks.fi

R. Lapatto
Hospital for Children and Adolescents, University of Helsinki, Helsinki, Finland

M. Nuutinen
Department of Pediatrics, University of Oulu, Oulu, Finland

Abbreviations

GH	Growth hormone
IGF-1	Insulin-like growth factor-1
LPI	Lysinuric protein intolerance
SLC7A7	Solute carrier family 7, member 7

Introduction

Lysinuric protein intolerance (LPI) is an autosomal recessive cationic amino acid transport defect (Perheentupa and Visakorpi 1965), which is caused by mutations in *SLC7A7* (Torrents et al. 1999). LPI is exceptionally common in Finland and almost half of the >100 reported patients are of Finnish origin (Simell 2001). The amino acid transporter y^{+}LAT-1 regulates intestinal absorption and renal loss of cationic amino acids (Simell 2001). Defect in the transporter leads to low plasma concentrations of lysine, arginine, and ornithine and the ensuing paucity of the urea cycle intermediates may increase the risk of hyperammonemia.

Treatment of LPI aims at prevention of hyperammonemia and optimization protein nutrition by protein restriction and supplementation with oral L-citrulline and lysine-HCl and, occasionally, with sodium benzoate or sodium phenylbutyrate (Simell 2001). Although current treatment regime combined with nutritional therapy effectively prevents acute episodes of hyperammonemia, the patients are at risk for malnutrition due to severe protein aversion. Consequently, subnormal growth is common in children with LPI.

Growth hormone (GH) promotes protein synthesis and accretion while reducing protein breakdown (Copeland and Nair 1994). It also stimulates food and protein intake (Roberts et al. 1995; Veyrat-Durebex et al. 1999; Blissett et al. 2000). GH action is mediated through insulin-like growth factor-1 (IGF-1) and hepatic IGF-1 synthesis is compromised in malnutrition (Palacio et al. 2002). There is a report on GH therapy in patients with organic acidemias with and without GH deficiency concluding that GH may be of value in some metabolic diseases (Marsden et al. 1994). We thus hypothesized that LPI patients might also benefit from GH therapy, but use of GH in metabolic diseases may also carry risks due to increased food consumption and subsequent potential overload of amino acids or other nutrients. To date, there is only one report of short-term GH therapy in LPI in a 12-year-old severely stunted girl with renal tubular disease as well (Esposito et al. 2006). To strengthen the proof of concept, we have treated four LPI children aged 7–16 years with recombinant GH for several years. We now report here the effects of GH therapy on growth and protein nutrition in these children.

Methods

Patient characteristics are presented in Table 1. All patients are Finnish and homozygous for the Finnish LPI founder mutation $LPI_{(Fin)}$ 1181-2A→T. Three of the patients had renal disease with poor growth, which is by itself an indication for GH therapy (Haffner et al. 2000). One patient was started on GH empirically to improve her growth.

Heights (to the nearest 0.1 cm) and weights (to the nearest 0.1 kg) were measured using Harpenden stadiometer and Soehnle S10 electronic scale, respectively. Yearly blood samples (for measurement of CBC, free T4, HbA_1c, IGF-1, NH_4, prealbumin, serum total and HDL-cholesterol and triglycerides, creatinine, cystatin-C, as well as plasma amino acids) were drawn after fasting, and the analyses were performed at the Central Laboratory of the Turku University Hospital using standardized methods. Insulin–arginine and clonidine provocation tests were performed using standard protocols. Bone age (Bayley–Pinneau) was estimated from the left hand and wrist. Estimation of dietary intakes was based on food records, which were checked for accuracy by a nutritionist.

Table 1 Patient characteristics during GH therapy

Patient no.	1	2	3	4
Sex	F	F	F	M
Age (year)	10.7	12.4	15.8	16.5
Medications				
Citrulline (mg/kg/day)	120	133	104	53
Lysine (mg/kg/day)	16	26	7	15
Leucine (mg/day)	–	–	1,000	–
Isoleucine (mg/day)	–	–	500	–
Sodium benzoate (mg/kg/day)	97	–	108	81
Sodium phenylbutyrate (mg/kg/day)	182	112	149	168
L-carnitine (g/day)	1	1	1	1
Calcium (mg/day)	500	1,500	500	1,000
Phosphate (mg/day)	–	1,000	–	–
Multivitamin (tabl/day)	1	1	1	1
Essential amino acid mixture (g/day)	15	–	–	12
Glucose polymer/corn starch (g/day)	100	25	100	–
Na-phosphate solution (Joulie, ml/day)	–	–	36	33
Citrate solution (Lightwood-I, ml/day)	–	–	150	–
Na hydrogenocarbonate (g/day)	–	2.5	–	9
Lovastatin (mg/day)	–	–	10	–
Other	–	–	Leuprorelin	–
Laboratory				
Creatinine (μmol/l)	34	79	131	95
Cystatine C (mg/l)	0.71	1.54	1.53	1.39
Creatinine clearance (ml/s/1.73 m^2)	–	0.46	0.65	–
HCO_3	–	20	20	18
Total cholesterol (mmol/l)	4.6	5.3	4.0	5.2
HDL cholesterol (mmol/l)	0.79	1.08	1.03	1.05
Triglycerides (mmol/l)	3.4	5.7	4.2	3.4

Results

IGF-1 Values, GH Provocative Tests and Effect of GH Therapy on Growth

IGF-1 values were low in all LPI patients before GH therapy (Table 2) but increased as expected during treatment. GH stimulation test was abnormal only in one of the patients and bone age was significantly delayed in all patients. Daily GH doses between 0.035 mg/kg and 0.050 mg/kg/day were used. Two of the patients (patients 1 and 2, both female) were prepubertal at the initiation of the GH therapy and remained so for years on GH therapy. Patient 3 (female) had entered puberty at the initiation of GH therapy, and she was treated with GnRH-analog (leuprorelin 75 mg/kg every 4 weeks) simultaneously with GH. Patient 4 (male) was just entering puberty (G2P1) at the initiation of the GH therapy.

During the 3–4.5 years of GH therapy, height standard deviation score (SDS) improved by 0.7–1.8 SDS (Fig. 1). First-year growth velocity increased in all but one patient (Patient 3), who was treated with GnRH-analog (see above). Bone age did not catch up during GH therapy, but IGF-1 values increased as expected (Table 2).

Nutrition and Dietary Intakes

Protein intake, in gram per kilogram per day, showed no clear differences during the GH therapy, although it increased when expressed in grams/day (Table 2). Oral citrulline and lysine doses showed a slight decreasing trend. Plasma arginine and lysine as well as prealbumin concentrations remained rather stable but plasma glutamine decreased markedly in three of the patients during GH therapy. There were no episodes of

Table 2 The effect of GH treatment in LPI patients

Patient no.	1	2	3	4
Sex	F	F	F	M
Age at introduction of GH therapy (year)	7.2	7.9	11.8	13.5
GH dose (mg/kg/day)	0.050	0.035	0.050	0.035
GH duration (year)	3.5 (ongoing)	4.5 (ongoing)	4 (ongoing)	3 (ongoing)
Antropometry	–	–	–	–
Birth weight (kg)	3.270	2.980	4.020	3.740
Birth length (cm)	50.5	48	52	52
Height (cm) at GH introduction	105.2	97.4	129.2	134.4
Height (SD) at GH introduction	−3.4	−5.5	−2.9	−2.7
Weight (kg) at GH introduction	15.8	14.4	26.9	29.7
Weight (%)[a] at GH introduction	−8	−5	+1	+1
Growth velocity 1 year prior to GH (cm/year)	4.8	4.0	7.4	4.2
Growth velocity during 1st year on GH (cm/year)	10.0	7.8	6.6	9.6
Current height (cm)	126.1	126.7	151.2	159.7
Current height (SD)	−2.1	−3.7	−2.1	−2.0
Laboratory	–	–	–	–
GH max (insulin-arginine-test, μg/l)	12.6	8.4	na	11.7
GH max (clonidine test, μg/l)	na	na	na	6.5
IGF-1 before GH treatment (nmol/l)	8	<5	11	13
IGF-1 during GH treatment (nmol/l)	10	11.6	40	49
Plasma arginine[b] before GH (μmol/l)	25	34	37	23
Plasma arginine during GH (μmol/l)	14	43	41	30
Plasma lysine[c] before GH (μmol/l)	82	74	144	89
Plasma lysine during GH (μmol/l)	65	120	141	71
Plasma glutamine[d] before GH (μmol/l)	2,181	1,682	1,331	963
Plasma glutamine during GH (μmol/l)	1,353	1,071	952	1,287
Urinary orotate before GH (μmol/mmol crea)	9.7	8.3	3.9	5.0
Urinary orotate during GH (μmol/mmol crea)	4.2	6.9	3.3	4.0
Plasma prealbumin before GH (mg/l)	166	na	207	270
Plasma prealbumin during GH (mg/l)	185	na	191	180
Bone age[e]	–	–	–	–
Before GH	−2 year	−3 year	−2.5 year	−4.5 year
After 2–3 years on GH	−2 year	−2.5 year	−2 year	−2.5 year
Nutrition and amino acid doses	–	–	–	–
Protein intake before GH (g/kg/day)	1.1	1.3	1.2	1.4
Protein intake during GH (g/kg/day)	1.1	1.2	1.2	na
Citrulline dose before GH (mg/kg/day)	132	182	130	99
Citrulline dose during GH (mg/kg/day)	120	133	104	53
Lysine dose before GH (mg/kg/day)	20	29	9	17
Lysine dose before GH (mg/kg/day)	16	26	7	15

[a]Percentual deviation from age and gender adjusted mean weight
[b]Reference range 23–86 μmol/l (from Dickenson et al. 1965)
[c]Reference range 71–151 μmol/l
[d]Reference range 57–467 μmol/l
[e]Bone age – calendar age

hyperammonemia or other adverse events. GH treatment was therefore well tolerated by all four LPI patients.

Discussion

Long-term recombinant GH therapy was effective and well tolerated in four LPI children aged 7–16 years of age as judged by improved growth and protein tolerance. The GH therapy was also safe as it was continued for more than 3 years without any apparent adverse events. Two of the female patients were prepubertal through many years on GH therapy but one female was pubertal at the GH onset. One male subject entered puberty at the start of GH treatment but, as the growth acceleration takes place in Tanner stage 3–4 in boys, his improved growth velocity was not due to pubertal growth spurt. Taken together, GH therapy offers new possibilities to improve growth in LPI. However, the effect of GH

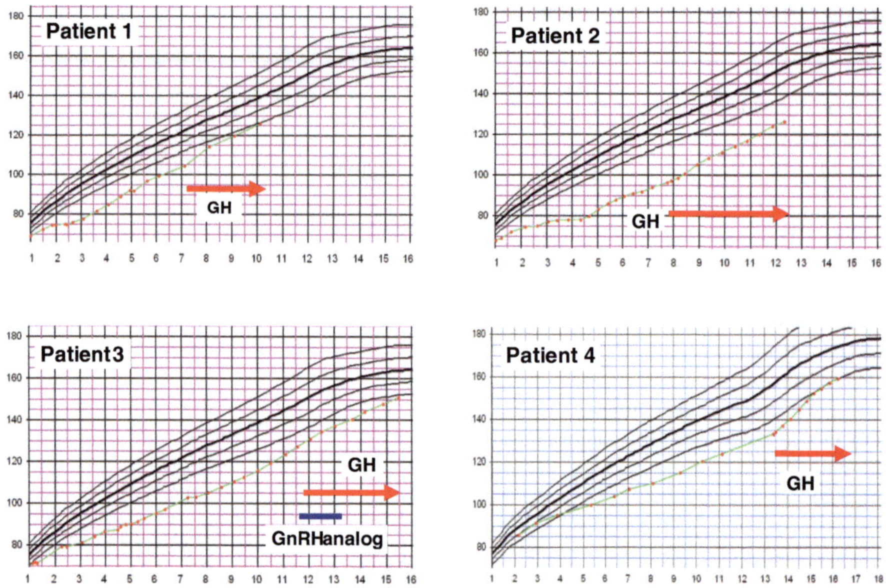

Fig. 1 Growth curves of four LPI patients with growth hormone therapy. *GH* growth hormone, *GnRH analog* gonadotropin-releasing hormone analog

on adult height achievement could not be measured as the patients have not yet reached adult height.

In LPI, spontaneous aversion to dietary protein develops at an early age. The LPI patients are encouraged to increase dietary protein intake modestly during citrulline and nitrogen-scavenging therapy, but aversion prevents many patients from accepting more than minimal requirements (Simell 2001). Children with LPI often show retarded growth and poor muscular development. Inadequate nutrition due to protein aversion may well contribute to the poor growth in LPI (Simell 2001) and some of the pathological manifestations of LPI may be related to deficiency of diamino acids (lysine, arginine, ornithine) and other essential amino acids, but the pathogenesis of the many problems associated with this disease is still poorly understood (Rajantie et al. 1980; Sidransky and Verney 1985; Simell 2001; Tanner et al. 2007). Compromised growth is not alleviated with citrulline supplementation. It may therefore be more related to reduced lysine availability than defective urea cycle function in LPI (Awrich et al. 1975; Tanner et al. 2007).

Interestingly, a recent study reported severe intrauterine growth restriction in LPI (*Slc7a7*-deficient) mice with markedly downregulated IGF-1 expression in fetal liver. Moreover, only a minority of LPI pups survived indicating clearly more severe phenotype in LPI mice than in humans (Sperandeo et al. 2007). In humans with LPI, intrauterine growth appears to be normal and growth failure is a postnatal phenomenon. The infants thrive as long as they are breastfed (Simell et al. 1975), whereas in mice the growth is retarded already in utero (Bröer 2007). In keeping with this, the LPI children in our study did not express intrauterine growth restriction, since their birth lengths varied between 48 and 52 cm and weights between 2,980 and 4,020 g.

Patients with LPI have decreased plasma levels of arginine, ornithine, and lysine. Intravenous arginine inhibits somatostatin release, and it is commonly used as GH-releasing stimulant in GH secretion tests. Oral arginine, when given in large doses, stimulates GH secretion (Collier et al. 2005), and oral arginine supplementation has been associated with improved growth in two children with LPI (Goto et al. 1984). Oral lysine ingestion acutely and markedly increases serum GH concentrations as well (van Vught et al. 2008). We use oral lysine supplementation in moderate doses to normalize plasma lysine values and oral citrulline to boost urea cycle. Oral arginine and large doses of oral lysine are not tolerated well in LPI since they may cause diarrhea and other abdominal complaints. Still, growth of many LPI patients remains compromised.

It is well known that GH promotes protein synthesis, protein accretion, and lipolysis and reduces protein breakdown (Copeland and Nair 1994). GH also stimulates food and protein intake (Roberts et al. 1995; Veyrat-Durebex et al. 1999; Blissett et al. 2000). GH/IGF axis is influenced by many factors, including malnutrition and renal disease, as recently reviewed by Richmond and Rogol (2008). In those situations, IGF-values are decreased whereas GH values are increased due to GH resistance (Grottoli et al. 2003). Many patients with inherited metabolic disease show impaired growth, but only few have been successfully alleviated with GH. It has, however, been used in patients with organic acidemias with and without GH deficiency but the value of it has remained unsure (Marsden et al. 1994). Prader–Willi children show increased anabolism and lean body mass and less fat mass gain if treated with GH (Eiholzer and Whitman 2004). Children with chronic renal failure show catch-up growth when given GH treatment and majority of them achieve normal adult height (Haffner et al. 2000). Previously, there is only one report of short-term GH therapy in a 12-year-old severely growth-retarded girl with LPI and tubular disease (Esposito et al. 2006). She had GH deficiency (peak GH secretion in a stimulation test was 6.4 μg/l) and presented a significant acceleration of growth from 2 cm/year before therapy to 8 cm/year during the first year on GH. Our four LPI patients showed analogous improvement of growth. Possible side effects of GH include slipped femoral epiphysis and benign cerebral hypertension; both of these might be more probable in LPI patients, who are prone to hyperammonemia-induced brain dysfunction and osteoporosis. However, neither of these complications was observed in our patients and the GH therapy was also otherwise well tolerated. In fact, decreased plasma glutamine concentrations may indicate a reduced risk of clinical hyperammonemia.

The weakness of this study is the low number of subjects and any decisive conclusions cannot be drawn. Moreover, two of the patients (one boy and one girl) were pubertal at the start of GH therapy. However, in boys the growth acceleration typically does not take place before Tanner stage 3, and the girl received also GnRH-therapy to delay the advance in bone age development.

In conclusion, patients with LPI show retarded growth, but GH therapy alleviated growth failure in 4 LPI children without measurable adverse effects on protein metabolism.

Acknowledgments This study was supported by the European Genomics Initiative on Disorders of Plasma Membrane Amino Acid Transporters (EUGINDAT), the Finnish Cultural Foundation, and the Novo Nordisk Foundation.

Take Home Message

Poor growth in children with lysinuric protein intolerance can be improved by growth hormone.

References

Awrich AE, Stackhouse WJ, Cantrell JE, Patterson JH, Rudman D (1975) Hyperdibasicaminoaciduria, hyperammonemia, and growth retardation: treatment with arginine, lysine and citrulline. J Pediatr 87:731–738

Blissett J, Harris G, Kirk J (2000) Effect of growth hormone therapy on feeding problems and food intake in children with growth disorders. Acta Paediatr 89:644–949

Bröer S (2007) Lysinuric protein intolerance: one gene, many problems. Am J Physiol Cell Physiol 293:C540–C541

Collier SR, Casey DP, Kanaley JA (2005) Growth hormone responses to varying doses of oral arginine. Growth Horm IGF Res 15: 136–139

Copeland KC, Nair KS (1994) Acute growth hormone effects on amino acid and lipid metabolism. J Clin Endocrinol Metab 78: 1040–1047

Dickenson JC, Rosenblum H, Hamilton PB (1965) Ion exchange chromatography of the free amino acids in the plasma of the newborn infant. Pediatrics 36:2–13

Eiholzer U, Whitman BY (2004) A comprehensive team approach to the management of patients with Prader-Willi syndrome. J Pediatr Endocrinol Metab 17:1153–1175

Esposito V, Lettiero T, Fecarotta S, Sebastio G, Parenti G, Salerno M (2006) Growth hormone deficiency in a patient with lysinuric protein intolerance. Eur J Pediatr 165:763–766

Goto I, Yoshimura T, Kuroiwa Y (1984) Growth hormone studies in lysinuric protein intolerance. Eur J Pediatr 141:240–242

Grottoli S, Gasco V, Ragazzoni F, Ghigo E (2003) Hormonal diagnosis of GH hypersecretory states. J Endocrinol Invest 26(10 suppl): 27–35

Haffner D, Schaefer F, Nissel R, Wuhl E, Tönshoff B, Mehls O (2000) Effect of growth hormone treatment on the adult height of children with chronic renal failure. German study group for growth hormone treatment in chronic renal failure. N Engl J Med 343:923–930

Marsden D, Barshop BA, Capistrano-Estrada S et al (1994) Anabolic effects of human growth hormone: management of inherited disorders of catabolic pathways. Biochem Med Metab Biol 52:145–154

Palacio AC, Pérez-Bravo F, Santos JL, Schlesinger L, Monckeberg F (2002) Leptin levels and IgF-binding proteins in malnourished children: effect of weight gain. Nutrition 18:17–19

Perheentupa J, Visakorpi J (1965) Protein intolerance with deficient transport of basic amino acids. Lancet 2:813–816

Rajantie J, Simell O, Perheentupa J (1980) Basolateral-membrane transport defect for lysine in lysinuric protein intolerance. Lancet 1: 1219–1221

Richmond EJ, Rogol AD (2008) Growth hormone deficiency in children. Pituitary 11:115–120

Roberts TJ, Azain MJ, White BD, Martin RJ (1995) Rats treated with somatotropin select diets higher in protein. J Nutr 125:2669–2678

Sidransky H, Verney E (1985) Chemical pathology of diamino acid deficiency: considerations in relation to lysinuric protein intolerance. J Exp Pathol 2:47–57

Simell O (2001) Lysinuric protein intolerance and other cationic aminoacidurias. In: Scriver CR, Beaudet AL, Sly WS, Valle D (eds) Childs B, Kinzler KW, Vogelstein B (assoc. eds) The metabolic and molecular bases of inherited disease, 8th edn. McGraw-Hill, New York, pp 4933–4956

Simell O, Perheentupa J, Rapola J, Visakorpi JK, Eskelin LE (1975) Lysinuric protein intolerance. Am J Med 59:229–240

Sperandeo MP, Annunziata P, Bozzato A et al (2007) Slc7a7 disruption causes fetal growth retardation by downregulating Igf1 in the mouse model of lysinuric protein intolerance. Am J Physiol Cell Physiol 293:C191–C198

Tanner LM, Näntö-Salonen K, Niinikoski H, Huoponen K, Simell O (2007) Long-term oral lysine supplementation in lysinuric protein intolerance. Metabolism 56:185–189

Torrents D, Mykkänen J, Pineda M et al (1999) Identification of SLC7A7, encoding y+LAT-1, as the lysinuric protein intolerance gene. Nat Genet 21:293–296

van Vught AJAH, Nieuwenhuizen AG, Brummer R-JM, Westerterp-Plantenga MS (2008) Effects of oral ingestion of amino acids and proteins on the somatotrophic axis. J Clin Endocrinol Metab 93:584–590

Veyrat-Durebex C, Gaudreau P, Coxam V, Gaumet N, Alliot J (1999) Peripheral injection of growth hormone stimulates protein intake in aged male and female rats. Am J Physiol 276:E1105–E1111

JIMD Reports
DOI: 10.1007/8904_2011_16

Received: 11 September 2009; Revised: 16 December 2010;
Accepted: 14 January 2011; Published online: 22 June
Communicated by: Nenad Blau

Outcomes of Phenylketonuria with Relevance to Follow-Up

F.J. van Spronsen and A. Bélanger-Quintana

Abstract Currently, there is no international consensus on how patients with phenylketonuria (PKU) (or milder forms of hyperphenylalaninaemia) should be followed in clinical practice. Guidelines concerning the frequency and type of assessments that should be made according to age usually focus on blood phenylalanine concentrations. A need exists for improved guidelines on how to do the follow-up of individuals with PKU/milder forms of hyperphenylalaninaemia. An interdisciplinary approach for monitoring patients is required, involving relevant clinical investigations and regular contact with a clinician and dietician/nutritionist as well as contact with social health worker, psychologist and neurologist, at least at request. This chapter presents a scheme for follow-up. However, by no means this scheme aims to present the one for all time follow-up programme. The scheme for follow-up may rather serve as a start for further discussion in larger groups of professionals in collaboration with patients and their parents. A number of questions remain unanswered, and further research is still needed to fine-tune the management of PKU at different ages.

Keywords Follow-up · Guidelines · Neurocognitive outcome · Nutritional deficiencies · Phenylketonuria · Phenylalanine

Abbreviations

IQ	Intelligence quotient
MHP	Mild hyperphenylalaninaemia
MRI	Magnetic resonance imaging
MRS	Magnetic resonance spectroscopy
PAH	Phenylalanine hydroxylase
Phe	Phenylalanine
PKU	Phenylketonuria
QoL	Quality of life

F.J. van Spronsen (✉)
Beatrix Children's Hospital, University Medical Center of Groningen, Groningen, The Netherlands
e-mail: f.j.van.spronsen@bkk.umcg.nl

A. Bélanger-Quintana
Unidad de Enfermedades Metabólicas, Servicio de Pediatría, Hospital Ramón y Caja, Madrid, Spain
e-mail: abelanger.hrc@salud.madrid.org

Introduction

Phenylketonuria (PKU; OMIM 261600) is an inherited autosomal recessive error in amino acid metabolism caused by a deficiency in the enzyme phenylalanine hydroxylase (PAH). Consequently, patients are unable to convert the essential amino acid phenylalanine (Phe) to tyrosine, with the result that plasma and tissue concentrations of Phe are elevated, negatively impacting on cognitive development and function (Scriver and Kaufman 2001). Untreated patients with the most severe form of PAH deficiency, usually termed 'classic' PKU, have Phe concentrations >1,200 μmol/L; less severe forms may be termed as 'mild PKU' (Phe 600–1,200 μmol/L) or 'mild hyperphenylalaninaemia' (MHP; Phe <600 μmol/L) (Blau et al. 2009). However, this classification is not that useful if untreated patients have not reached their maximal Phe concentration at the start of treatment due to early diagnosis with neonatal screening. In this case, the use of other strategies such as genotyping or the determination of dietary Phe tolerance is necessary (Guldberg et al. 1998; van Spronsen et al. 2009a).

Screening for high concentrations of Phe in newborns by heel puncture is crucial for identifying PKU patients, so that treatment can be implemented early to prevent severe mental retardation and allow optimal cognitive development. Prevention of mental retardation has been the primary aim of PKU treatment for a long time (van Spronsen and Burgard 2008), but even among patients with well-controlled Phe concentrations, intellectual outcome is not completely normal (Feillet et al. 2010). Neurocognitive deficits may include

Table 1 Targets of blood phenylalanine concentrations and frequencies for control for different age groups[a]

Issue/age in years	0–1	1–4	4–10	10–12	12–16	>16
Frequency blood phenylalanine/month[b]	4 (1–6)	3 (1–4)	1.5 (1–3)	1 (1–2)	1 (0.5–2)	1 (0.33–2)
Frequency blood amino acid measurements/year[c]	2 (0–4)	1 (0–4)	0 (0–2)	1 (0–1)	1 (0–1)	1 (0–2)
Frequency clinical evaluation/year[c]	9 (2–12)	4 (1–6)	3 (1–4)	2 (1–4)	1 (1–3)	1 (1–2)
Range of target phenylalanine (μmol/L)[d]	120–360 (<400)	120–360 (<400)	120–400 (<480)	120–360 (<900)	120–600 (<900)	120–700 (<900)

[a]Adapted from van Spronsen et al. (2009b)
[b]Most frequent observed number of blood phenylalanine/amino acid concentration measurement; the complete reported range is given within brackets
[c]Most frequent number of clinical evaluations; the complete reported ranges are given within brackets
[d]Most frequent reported target of phenylalanine concentrations; the complete reported ranges are given within brackets

deficits in executive function tasks such as organization, inhibitory control and planning, which may be responsible for poor levels of academic achievement (Huijbregts et al. 2002; Feldmann et al. 2005; Gassió et al. 2005).

A strong relationship has been established between the control of blood Phe concentrations during infancy and childhood and intelligence quotient (IQ) (Scriver and Kaufman 2001). Therefore, good control of Phe concentrations has been the main objective of treatment, but recommended maximum Phe blood concentrations by age group vary considerably in all age groups in Europe, as do frequencies of monitoring (Table 1).

Only some countries have internationally available reported national guidelines. These guidelines focus on monitoring target Phe concentrations, frequency of blood sampling and clinical evaluations but usually provide inadequate information for the most appropriate follow-up of patients with PKU at different ages (van Spronsen and Burgard 2008). These guidelines, for example, seldom offer advice regarding nutritional issues or type and timing of appropriate neurocognitive tests.

When designing any kind of clinical algorithm for the follow-up of a patient with PKU, it is therefore important to be sure of the objective. Where in the past 'normal cognitive function' was the primary aim, at present 'a life as normal as possible' is the main target for many clinicians (van Spronsen and Burgard 2008), aiming not only at normal neuropsychological test outcomes, but also at normal qualities of life. When aiming at a day-to-day life as normal as possible, target outcomes need to be re-identified. This chapter presents a scheme for follow-up with issues that need attention and consideration for simple or more extensive monitoring (Table 2). This scheme may serve as a start for further discussion to achieve guidelines rather than aiming to present the one for all time follow-up programme. Guidelines need to be developed by larger groups of physicians, psychologists, nutritionists, biochemists and other professionals in cooperation with patients and their parents.

Designing a PKU Follow-Up Programme: Points to Consider

Individually Tailored Treatment

The principal question whether all patients need the same strictness of treatment is unanswered. There is even a clear variation among different countries in the Phe concentrations that determine the beginning of treatment in patients with a positive neonatal screening result (Blau et al. 2010; Feillet et al. 2010). Old reports already claimed that some untreated patients did not develop severe mental retardation (Pitt and Danks 1991). The work of different research groups using magnetic resonance spectroscopy (MRS) have brought this discussion alive again by detecting different brain Phe levels in patients with the same blood Phe concentrations (Weglage et al. 2001; Pietz et al. 2002).

In other diseases such as leukaemia, the type of presentation, biochemical and molecular parameters help to distinguish patients who need a different degree of treatment (Faderl et al. 2010). One would expect that if treatment strategies can be based on molecular data in leukaemia, it should be possible in monogenetic diseases such as PKU with a simple monogenic trait. Studies with DNA have claimed to distinguish patients with more and less severe degrees of PAH deficiency, but at the same time are not always conclusive (Kayaalp et al. 1997). However, it was already in 1999 that Scriver et al. presented ideas why there is not such a simple trait in PKU (Scriver and Waters 1999), elaborating these concepts further in one of his later papers (Scriver 2007). So far, in PKU we have been unable to identify markers that would enable different treatment strategies. Clearly, complex studies in large populations are needed before we can answer to a sufficient degree the question as to which patients need treatment and with which strictness and which do not. Until this question is safely answered, all patients should be offered the same treatment strategy in relation to the therapeutic target Phe concentrations.

Table 2 Issues of monitoring in standard and expert follow-up in PKU patients

	Standard PKU follow-up	Expert PKU follow-up: differences to standard follow-up
Clinical and nutritional follow-up including visit to paediatrician/physician metabolic disease and dietician metabolic diseases with measures of growth and development	0–12 months: monthly (after parents are accustomed to PKU and its treatment) 1–4 years: every 3 months 4–18 years: every 6 months Adults: every 6–12 months Pregnancy: monthly	
Phenylalanine concentration monitoring	0–12 months: weekly 1–2 years: twice monthly >2 years: monthly Pregnancy: weekly	0–24 months: weekly 2–4 years: twice monthly >4 years: once to twice monthly Pregnancy: twice weekly
Individual adaptation of care		Analysis of DNA and tolerance of phenylalanine with possible consequences for strictness of treatment Investigation of possibility of other treatment strategies for individual including tetrahydrobiopterin
Nutritional investigations	*Yearly*: Haemoglobin, leucocytes, thrombocytes Calcium, phosphate, alkalic phosphatase, ALT	*Yearly:* Blood: holotranscobalamine, methylmalonic acid, homocysteine, total amino acids, pre-albumin, transferrine Urine: calcium, protein, phosphate, creatinine *Every 5 years*: Blood: carnitine, fatty acid profile, zinc, selenium Bone densiometry each 5–10 years starting at 15–20 years of age. Higher frequency when indicated. When abnormal, consider further investigations including markers of bone turnover
Neurological follow-up	Regular clinical evaluations, special attention for tremor, brisk reflexes Specific neurological tests by paediatric/adult neurologist when indicated by the results of paediatrician/physician metabolic disease	MRI with DTI if necessary (MRS is research rather than clinical follow-up)
Neurocognitive function	Regular school reports with special attention for attention, hyperactivity Determination IQ once between 6 and 8 years of age	Response speed test BRIEF if possible
Psychosocial issues and quality of life		Tests need to be developed

BRIEF Behaviour Rating Inventory of Executive Function, *DTI* diffusion tensor imaging

Normal Neurological Outcomes

A strong relationship has been established between the control of blood Phe concentrations during infancy and childhood and IQ. Waisbren et al. (2007) conducted a meta-analysis of 40 studies, which pointed out that in early treated PKU patients, an incremental rise in Phe of 100 μmol/L was predictive of a decline in IQ between 1.3 and 3.1 points at Phe concentrations >423 μmol/L during critical periods from 0 to 12 years of age. However, the relation between IQ and blood Phe concentrations may be more complex. For example, in a study of 46 patients, Anastasoaie et al. (2008) found that IQ tended to relate to the fluctuation of the Phe concentration rather than to the Phe concentrations being above a specific threshold. A recent meta-analysis by Albrecht et al. (2009) demonstrated the negative effects of concurrent Phe concentrations on reaction times among children at levels >250 μmol/L, and >600 μmol/L in adolescents. No clear cut-off value could be estimated for adults, as there is little data relating Phe concentrations and brain function in this age group.

Although well-controlled patients reach a normal IQ, it may be lower than their siblings or classmates. Above this, patients with normal IQ still might have milder neurocognitive deficits such as slow executive function, difficulties in organization and planning, etc. (Huijbregts et al. 2002; Feldmann et al. 2005; Gassió et al. 2005). Most present guidelines do not consider the routine investigation of these functions of which deficits may be minor but at the same time may be responsible for day-to-day difficulties in the patients and in the end may unnecessarily lead to poor levels of academic achievement in some of the patients.

Notwithstanding the relationship between blood Phe concentrations and outcome, it is to be resolved whether blood Phe concentrations can fully explain the total pathogenesis of brain pathophysiology in PKU (van Spronsen et al. 2009c). Hoeksma et al. (2009) evaluated the cerebral protein synthesis rate in relation to plasma Phe concentration in adult patients with PKU, using positron emission tomography after the administration of intravenous l-[1-^{11}C]-tyrosine. A significant negative relationship (R^2=0.40, $p < 0.01$) between plasma Phe concentration and cerebral protein synthesis rate was observed in 16 patients with PKU. Individuals with Phe concentrations >600–800 μmol/L experienced a greater decrease in cerebral protein synthesis at these higher plasma Phe concentrations, compared with lower Phe concentrations.

Psychosocial Outcomes

An additional target of significance among patients with PKU includes psychosocial outcome, of which too few studies have been conducted to date. Patients with PKU may suffer severe psychological problems, such as depression, agoraphobia and low self-esteem (Hendrikx et al. 1994; Ris et al. 1994; Sullivan and Chang 1999; Waisbren et al. 1994; Waisbren and Levy 1991; Weglage et al. 1992; Weglage et al. 1994), and improved dietary control may assist with improving these psychological outcomes.

Pietz et al. (1997) assessed psychiatric disorders in patients with PKU and tested whether biochemical control, intellectual functioning, white matter abnormalities visible on magnetic resonance imaging (MRI) and/or style of parenting was related to psychopathology. Findings indicated that a higher percentage of patients with PKU internalize their feelings (25.7% of patients with PKU versus 8.3% of the control group), which are later expressed as psychiatric problems of anxiety, depression and low self-esteem (Pietz et al. 1997). In a retrospective study conducted by Weglage et al. (1992), 34 early treated normally intelligent adolescents with PKU and their parents were tested using several psychometric personality inventories and self-developed questionnaires to assess their psychosocial situation, disease- and diet-specific knowledge. Compared to the control population, patients had a more negative evaluation of their scholastic ability, were less motivated, had a lower tolerance of frustration and gave more negative self-descriptions. They also tended to be less extroverted and had a higher level of dependency on their families.

Nutritional Outcomes

A further main target is to ensure that there are no nutritional deficiencies that may give rise to possible physiological or neurological concerns. These nutritional deficiencies cannot be monitored by simply measuring Phe concentrations. Some guidelines advice concerning a prescriptive total amount of protein, as well as vitamin B_{12} monitoring (Medical Research Council for Phenylketonuria 1993; Abadie et al. 2005). Research papers have studied the possibility and clinical importance of deficiencies of amino acids other than Phe, bone turnover markers (such as deoxypyridinoline, osteocalcin, bone alkaline phosphatase, C-terminal procollagen peptide type I, osteoprotegerin), calcium, carnitine, coenzyme Q_{10}, folates, iron, long-chain polyunsaturated fatty acids, selenium, vitamin A, B_2 and B_{12}, C and E, and zinc (Ambroszkiewicz et al. 2004; Millet et al. 2005; Feillet and Agostini 2010). Effects on bone density, growth and oxidative stress – with possible consequences for intellectual development and neuropsychological function – have been suggested although clear evidence is lacking (Sirtori et al. 2005; Feillet and Agostini 2010).

Until now, there have been no studies that specify in which situations patients are more prone to specific nutritional deficiencies. Is it related to age or growth velocity? Is it related to how strict their diet is, or rather, is it more frequent in patients who relax their diet but still do not have a quantitatively and/or qualitatively normal protein intake? It also remains to be investigated to what degree subclinical nutritional deficits may impair optimal neuropsychological and psychosocial function. It is still unclear, therefore, at what age patients would benefit the most from arranging specific tests to detect nutritional deficits.

It is our opinion that there are probably three moments in life in which nutritional deficiencies of the patient are more prone to appear and therefore patients should be especially monitored: young infants, adolescents and elderly. Young infants have a rapid growth and development. In this age group, nutritional care may focus at deficiencies of essential amino acids and long-chain polyunsaturated fatty acids. Adolescents and adults, on the other hand, may abandon their restrictive diet, think they have normalized their natural protein intake and believe they do not need their amino acid supplements anymore. Such patients tend to be nutritionally deficient, as their intake of natural protein is both of low quantity and quality. In this age, group tests are needed to detect specific nutritional deficits such as vitamin B_{12}, (measuring methylmalonic acid and homocysteine rather than vitamin B_{12} itself) (Vugteveen et al. 2010) which may result in a less optimal physical condition and neurological dysfunction. With regard to elderly, there simply is not enough experience, but deficiencies may have a remarkable resemblance compared to young infants.

Pregnancy is another delicate situation in which dietary restrictions might need to be intensified for the sake of the foetus. Notwithstanding the fact that outcome is very good when Phe restriction is started clearly before conception and continued throughout pregnancy, there are nutritional issues. First of all, it sometimes proves difficult to prevent Phe concentrations from being too low in the second half of pregnancy when there is a high need of Phe due to growth velocity of both child and placenta. Tyrosine is an ongoing subject for debate. Should tyrosine be supplemented during pregnancy and how should the amount of supplementation be determined? There is a need of long-chain polyunsaturated fatty acids. Carnitine concentrations can be very low, but whether supplementation is needed is unclear. With regard to vitamins, there is the risk of vitamin A toxicity rather than deficiency.

Quality of Life Outcomes

As with the lack of PKU studies assessing psychosocial outcomes, there is also a dearth in the number of studies that assess quality of life (QoL) in patients with PKU. Two of the ones that have done so, however, display conflicting outcomes. Bosch et al. (2007) conducted a study to assess the course of life, socio-demographic outcomes and health-related QoL in 32 young adult patients (aged 18–30 years) with PKU who remained on treatment. Findings from the completed Course of Life questionnaire (Grootenhuis et al. 2003), the RAND-36 Health Survey and the cognitive scale of the TNO-AZL Adult Quality of Life questionnaire (Fekkes et al. 2000) were comparable to controls except for the higher percentage of patients requiring special education in primary school; this was, however, comparable with that of the patients' peers. The precise meaning of this finding for treatment strategies remains unanswered. In conclusion, Bosch et al. (2007) surmised that while PKU is a chronic disease with the burden of strict dietary control, early and continuously treated patients with PKU can enjoy a normal health-related QoL.

By contrast, Simon et al. (2008) showed that 47% of 67 young adult patients with PKU in Germany lived with their parents (compared with 25% in the general population), and more than 75% of male patients were not involved in a steady relationship. In addition, 9% of female and 18% of male patients had children, compared with approximately 50% of subjects in the general population in the same age group. Therefore, the data of Simon et al. (2008) substantiate the findings of another German study of Weglage et al. (1992) concerning decreased autonomy, and by that show that in PKU psychosocial issues still need to be addressed to achieve normal QoL outcome.

New studies will have to reveal whether the differences in data of Bosch et al. (2007) compared to Simon et al. (2008) and Weglage et al. (1992) are due to differences in study design and/or study groups.

To date, there is no study correlating psychosocial and QoL issues with levels of blood Phe, and the stringency of Phe dietary restriction to achieve these Phe concentrations. Furthermore, little data on the influence of family issues and style of parenting on QoL are available. We also have no data about the relation between neuropsychological and psychosocial outcomes at low(er) and high(er) Phe concentrations (during various periods of life), and the effort it takes for patients to achieve these concentrations. Therefore, we cannot be sure of the relationship of cause and effect. It might very well be that, on the one hand, stringent dietary restrictions resulting in well-controlled Phe concentrations allow the best possible neurocognitive outcomes in the tests applied, but that, on the other hand, such a stringent dietary restriction may imply a restriction of social life and psychosocial well-being that results in the incapability of patients to completely use their biological potentials of neurocognitive skills in day-to-day life. At this moment, it is unclear whether a lower than optimal control of Phe concentrations, the stringent dietary restrictions or a combination of both negatively influence self-esteem in PKU patients (Hendrikx et al. 1994; Weglage et al. 1992), and consequently result in less autonomy (Weglage et al. 1992; Simon et al. 2008).

Crone et al. (2005) reported that too a strict handling of the dietary limitations is not necessarily associated with lower Phe concentrations. Therefore, teaching coping strategies to our patients and their families may be of more importance than considered previously, not only at the moment of diagnosis but also at all other age groups, such as adolescence and adulthood.

Designing a PKU Follow-Up Programme: What Elements Should It Contain?

Table 2 contains the elements that we believe need attention in the follow-up of PKU patients. A PKU centre is the likely venue for this follow-up. These centres should always have a metabolic paediatrician/adult physician, a dietician or nutritionist, and a clinical biochemist. Other staff who ideally should be included in this interdisciplinary team is a social health worker, a psychologist, a neurologist and a specialized nurse. One study showed that centralized expert teams provide better care than smaller centres (Camfield et al. 2004). However, the question remains as to what specifications distinguish PKU expert centres from teams who provide more standard care to their patients, who should decide on this issue and on what basis. At present, the suggested 'expert' follow-up does not include specific neuropsychological test batteries, MRS, positron emission tomography and extensive questionnaires on the QoL, as these are considered to be research rather than expert care.

Phe concentrations are still a key element in the monitoring of PKU patients, but other biochemical parameters should be considered to be able to assess their importance in the future, such as the fluctuation of these concentrations and the ratio of Phe to tyrosine and the other amino acids. Patients of all ages and phenotypes (on strict diet or not) deserve nutritional monitoring to avoid nutritional deficiencies. The strictness of their diet, age and personal situation should be taken into account to determine specific parameters.

Patients should be routinely followed to assess neurological outcome. A clinical neurological examination can be done by any experienced paediatrician/adult physician. Referral to a neurologist is advised in cases in which abnormalities are found. Additional neurological imaging (e.g. cerebral MRI) is encouraged in such cases. It should be remembered, however, that the relation between an abnormal MRI and clinical outcome is not that strong. Although a better control should try to be obtained, abnormalities on a MRI should not be used to impress the patient and by this means try to improve the adherence to treatment. When using MRI, the use of more advanced techniques such as diffusion tensor imaging is advocated. At this moment, the use of MRS and positron emission tomography studies is for research rather than day-to-day follow-up.

IQ testing is considered appropriate for research and routine purposes at least until the age of 7. We encourage the use of neuropsychological tests (especially in patients over 7 years of age) to assess neuropsychological deficits such as executive function deficits. It is preferable to use standardized batteries of tests to achieve a greater degree of certainty. In this light, it is of importance to note that simple speed tests may be of practical value (Albrecht et al. 2009), whereas rather more complex neuropsychological batteries are needed for research. The Behaviour Rating Inventory of Executive Function (BRIEF) for paediatric ages and BRIEF-A for adults, a questionnaire, can be used but availability is especially in English and Spanish language (Waisbren and White 2010).

Measuring psychosocial outcome has become of more importance. A simple standardized questionnaire that could be used in a clinical setting would be of great help to understand the needs of the patients in different moments of their lives. It would be of added value when the QoL of patients and their families could be assessed with a simple standardized questionnaire as well. The report of Waisbren and White (2010) is the first one to present some ideas about such tests including the Behaviour Assessment System for Children (BASC-II), the Beck Anxiety Inventory 2nd edition for adults (BAI-II) and the Beck Depression Inventory 2nd edition (BDI-II) for adults. However, the usefulness of the BASC-II, the BAI-II and the BDI-II has to be proven in PKU patients, and therefore, research is necessary before these tests can be considered daily clinical practice for either standard or specialized care.

Conclusion

Notwithstanding the large library of knowledge in PKU, there is still no clear picture of how a follow-up programme can be successfully delivered. Once an individual has been diagnosed with PKU and is undergoing treatment, he or she deserves adequate follow-up during his or her entire lifespan, with differences as personal circumstances change at different ages. Each follow-up programme should be in the best interest of the patients with regards to their disease severity, age and other relevant factors, such as the desire to become pregnant.

This chapter tries to present a start for a follow-up programme. By no means, however, this chapter is aimed to present the one for all time follow-up programme. By giving some specific ideas on the programme, it may serve as a start for further discussion. Metabolic physicians, nutritionists, psychologists, biochemists, social health workers and specialized nurses are needed to achieve guidelines, in collaboration with parents and patients.

Competing Interests

Francjan van Spronsen and Amaya Bélanger-Quintana are both members of the scientific advisory board of Merck Serono S.A. – Geneva, Switzerland (an affiliate of Merck KGaA, Darmstadt, Germany) regarding treatment for phenylketonuria. Francjan J van Spronsen received grants and honoraria from Milupa, Scientific Hospital Supplies, Nutricia, and Merck Serono S.A. Amaya Bélanger-Quintana received grants from Mead-Johnson and Merck Serono S.A.

References

Abadie V, Berthelot J, Feillet F et al (2005) Association française pour le dépistage et la prevention des handicaps de l'enfant (AFDPHE). Arch Pediatr 12:594–601

Albrecht J, Garbade SF, Burgard P (2009) Neuropsychological speed tests and blood phenylalanine levels in patients with phenylketonuria: a meta-analysis. Neurosci Biobehav Rev 33:414–421

Ambroszkiewicz J, Gajewska J, Laskowska-Klita T (2004) A study of bone turnover markers in prepubertal children with phenylketonuria. Eur J Pediatr 163:177–178

Anastasoaie V, Kurzius L, Forbes P, Waisbren S (2008) Stability of blood phenylalanine levels and IQ in children with phenylketonuria. Mol Genet Metab 95:17–20

Blau N, Bélanger-Quintana A, Demirkol M et al (2009) Optimizing the use of sapropterin (BH4) in the management of phenylketonuria. Mol Genet Metab 96:158–163

Blau N, van Spronsen FJ, Levy HL (2010) Phenylketonuria. Lancet 376:1417–1427

Bosch AM, Tybout W, van Spronsen FJ, de Valk HW, Wijburg FA, Grootenhuis MA (2007) The course of life and quality of life of early and continuously treated Dutch patients with phenylketonuria. J Inherit Metab Dis 30:29–34

Camfield CS, Joseph M, Hurley T, Campbell K, Sanderson S, Camfield PR (2004) Optimal management of phenylketonuria: a centralized expert team is more successful than a decentralized model of care. J Pediatr 145:53–57

Crone MR, van Spronsen FJ, Oudshoorn K, Bekhof J, van Rijn G, Verkerk PH (2005) Behavioural factors related to metabolic control in patients with phenylketonuria. J Inherit Metab Dis 28:627–637

Faderl S, O'Brien S, Pui CH et al (2010) Adult acute lymphoblastic leukemia: concepts and strategies. Cancer 116:1165–1176

Feillet F, Agostini C (2010) Nutritional issues in treating phenylketonuria. J Inherit Metab Dis 33:659–664, Epub Feb 2010

Feillet F, van Spronsen FJ, Macdonald A et al (2010) Challenges and pitfalls in the management of phenylketonuria. Pediatrics 126:233–241

Fekkes M, Theunissen NC, Brugman E et al (2000) Development and psychometric evaluation of the TAPQOL: a health-related quality of life instrument for 1-5-year-old children. Qual Life Res 9:961–972

Feldmann R, Denecke J, Grenzebach M, Weglage J (2005) Frontal lobe-dependent functions in treated phenylketonuria: blood phenylalanine concentrations and long-term deficits in adolescents and young adults. J Inherit Metab Dis 28:445–455

Gassió R, Artuch R, Vilaseca MA et al (2005) Cognitive functions in classic phenylketonuria and mild hyperphenylalaninaemia: experience in a paediatric population. Dev Med Child Neurol 47: 443–448

Grootenhuis MA, Stam H, Destree-Vonk A, Heijmans HSA, Last BF (2003) Levensloop vragenlijst voor jong-volwassenen (Course of life questionnaire for Young adults). Gedrag en Gezondheid 31:336–350

Guldberg P, Rey F, Zschocke J, Romano V, François B, Michiels L et al (1998) A European multicenter study of phenylalanine hydroxylase deficiency: classification of 105 mutations and a general system for genotype-based prediction of metabolic phenotype. Am J Hum Genet 63:71–79

Hendrikx MM, van der Schot LW, Slijper FM, Huisman J, Kalverboer AF (1994) Phenylketonuria and some aspects of emotional development. Eur J Pediatr 153:832–835

Hoeksma M, Reijngoud DJ, Pruim J, de Valk HW, Paans AM, van Spronsen FJ (2009) Phenylketonuria: high plasma phenylalanine decreases cerebral protein synthesis. Mol Genet Metab 96:177–182

Huijbregts SC, de Sonneville LM, Licht R, van Spronsen FJ, Verkerk PH, Sergeant JA (2002) Sustained attention and inhibition of cognitive interference in treated phenylketonuria: associations with concurrent and lifetime phenylalanine concentrations. Neuropsychologia 40:7–15

Kayaalp E, Treacy E, Waters PJ, Byck S, Nowacki P, Scriver CR (1997) Human phenylalanine hydroxylase mutations and hyperphenylalaninemia phenotypes: a metanalysis of genotype-phenotype correlations. Am J Hum Genet 61:1309–1317

Medical Research Council for Phenylketonuria (1993) Recommendations on the dietary management of phenylketonuria. Report of Medical Research Council Working Party on phenylketonuria. Arch Dis Child 68:426–427

Millet P, Vilaseca MA, Valls C et al (2005) Is deoxypyridinoline a good resorption markerto detect osteopenia in phenylketonuria? Clin Biochem 38:1127–1132

Pietz J, Fätkenheuer B, Burgard P, Armbruster M, Esser G, Schmidt H (1997) Psychiatric disorders in adult patients with early-treated phenylketonuria. Pediatrics 99:345–350

Pietz J, Rupp A, Burgard P, Boesch C, Kreis R (2002) No evidence for individual blood-brain barrier phenylalanine transport to influence clinical outcome in typical phenylketonuria patients. Ann Neurol 52:382–383

Pitt DB, Danks DM (1991) The natural history of untreated phenylketonuria over 20 years. J Paediatr Child Health 27:189–190

Ris MD, Williams SE, Hunt MM, Berry HK, Leslie N (1994) Early-treated phenylketonuria:adult neuropsychologic outcome. J Pediatr 124:388–392

Scriver CR (2007) The PAH gene, phenylketonuria, and a paradigm shift. Hum Mutat 28:831–845

Scriver CR, Kaufman S (2001) Hyperphenylalaninemia: phenylalanine hydroxylase deficiency. In: Scriver CR, Beaudet AL, Sly WS, Valle D, Childs B, Vogelstein B (eds) The metabolic and molecular bases of inherited disease, 8th edn. McGraw-Hill, New York, pp 1667–1724

Scriver CR, Waters PJ (1999) Monogenic traits are not simple: lessons from phenylketonuria. Trends Genet 15:267–272

Simon E, Schwarz M, Roos J et al (2008) Evaluation of quality of life and description of the sociodemographic state in adolescent and young adult patients with phenylketonuria (PKU). Health Qual Life Outcomes 26:25

Sirtori LR, Dutra-Filho CS et al (2005) Oxidative stress in patients with phenylketonuria. Biochim Biophys Acta 1740:68–73

Sullivan JE, Chang P (1999) Review: emotional and behavioural functioning in phenylketonuria. J Pediatr Psychol 24:281–299

van Spronsen FJ, Burgard P (2008) The truth of treating patients with phenylketonuria after childhood: the need for a new guideline. J Inherit Metab Dis 31:673–679

van Spronsen FJ, van Rijn M, Dorgelo B et al (2009a) Phenylalanine tolerance can already reliably be assessed at the age of 2 years in patients with PKU. J Inherit Metab Dis 32:27–31

van Spronsen FJ, Ahring KK, Gizewska M (2009b) PKU – What is daily practice in various centres in Europe? J Inherit Metab Dis 32:58–64

van Spronsen FJ, Hoeksma M, Reijngoud DJ (2009c) Brain dysfunction in phenylketonuria: is phenylalanine toxicity the only possible cause? J Inherit Metab Dis 32:46–51

Vugteveen I, Hoeksma M, Bjorke Monsem AL et al (2010) Serum vitamin B_{12} concentrations within reference values do not exclude functional vitamin B_{12} deficiency in PKU patients of various ages. Mol Genet Metab 102:13–17, Epub October 2010

Waisbren SE, Levy HL (1991) Agoraphobia in phenylketonuria. J Inherit Metab Dis 14:755–764

Waisbren S, White DA (2010) Screening for cognitive and social-emotional problems in individuals with PKU: tools for use in the metabolic clinic. Mol Genet Metab 99(Suppl 1):S96–S99

Waisbren SE, Brown MJ, de Sonneville LM, Levy HL (1994) Review of neuropsychological functioning in treated phenylketonuria: an information processing approach. Acta Paediatr 407:98–103

Waisbren SE, Noel K, Fahrbach K et al (2007) Phenylalanine blood levels and clinical outcomes in phenylketonuria: a systematic literature review and meta-analysis. Mol Genet Metab 92:63–70

Weglage J, Fünders B, Wilken B et al (1992) Psychological and social findings in adolescents with phenylketonuria. Eur J Pediatr 151:522–525

Weglage J, Rupp A, Schmidt E (1994) Personality characteristics in patients with phenylketonuria treated early. Pediatr Res 35:611–613

Weglage J, Wiedermann D, Denecke J et al (2001) Individual blood-brain barrier phenylalanine transport determines clinical outcome in phenylketonuria. Ann Neurol 50(463):467

JIMD Reports
DOI: 10.1007/8904_2011_17

Received: 20 September 2010; Revised: 5 December 2010;
Accepted: 14 December 2010; Published online: 22 June
Communicated by: Ed Wraith

Three Adult Siblings with Mucopolysaccharidosis Type II (Hunter Syndrome): A Report on Clinical Heterogeneity and 12 Months of Therapy with Idursulfase

Michel C. Tchan, Kerry T. Devine, and David O. Sillence

Abstract Mucopolysaccharidosis type II (MPS II – Hunter syndrome) is a rare X-linked recessive disease of lysosomal glycosaminoglycan metabolism leading to a systemic storage disorder. We report three adult brothers (aged 46–52 years) with attenuated Hunter syndrome after 12 months of enzyme replacement therapy with idursulfase. Before enzyme replacement therapy, each had serious complications of their disease: in addition to all requiring urgent cervical spinal canal decompressions in middle age, one required emergency aortic and mitral valve surgery, another had stage IV airways disease, and the third had acute glaucoma resulting in unilateral blindness. One brother discontinued therapy after 12 months. The other two brothers reported subjective improvements in energy and exercise tolerance.

Keywords Enzyme replacement therapy · Hunter syndrome · Mucopolysaccharidosis type II

Introduction

Mucopolysaccharidosis type II (OMIM #309900) is an X-linked recessive disorder due to deficiency of iduronate-2-sulfatase (*IDS*, EC 3.1.6.13) which is responsible for the cleavage of O-linked sulfates from dermatan sulfate and heparan sulfate. The incidence is approximately 1.3 per 100,000 live male births (Poorthuis et al. 1999; Baehner et al. 2005), although affected females with skewed X-inactivation have been described (Manara et al. 2010). Hunter syndrome demonstrates a spectrum of disease with severe patients defined by cognitive impairment and progression of somatic disease with death from cardiac, respiratory, or neurological events in the second decade of life. Patients with the attenuated form do not have cognitive involvement and survive to adulthood, but have variable and often severe somatic involvement including cardiac valvular disease, airway compromise, spinal canal stenosis, short stature, joint restriction, dysostosis multiplex, hepatomegaly, and carpal tunnel syndrome (Wraith et al. 2008a, b).

Enzyme replacement therapy with idursulfase for MPS II was made available in Australia in 2008 for patients with the nonneurological form of the disease under the Federal Department of Health Life Saving Drugs Program. Idursulfase (Elaprase – Shire Human Genetic Therapies) is a recombinant human protein produced in HT1080 human fibroblast cultured cells that is posttranslationally modified to form active enzyme (Muenzer et al. 2006, 2007). Clinical trials demonstrated reduction in urinary glycosaminoglycan (GAG) excretion, an increase in the 6-min walk test (6MWT) distance, improvement in forced vital capacity (FVC), and normalization of organomegaly (Muenzer et al. 2006, 2007).

Adult siblings with MPS II have not, to our knowledge, been described in the literature. All these three men had symptoms in early childhood but were not diagnosed until young adulthood. The causative family mutation in *IDS* was determined in 2002 and reported as p.S152G. Two of the brothers have their own children, and one of them has a grandson who was determined to be unaffected by prenatal diagnosis (Fig. 1). Before treatment with idursulfase, at 0.5 mg/kg/week, they all had significant skeletal and connective tissue disease and had required cervical laminectomy for severe symptomatic cervical spinal canal stenosis as well as carpal tunnel releases. Idursulfase was commenced

M.C. Tchan (✉) and D.O. Sillence
Department of Genetic Medicine, Westmead Hospital, Westmead, NSW, Australia
and
Discipline of Genetic Medicine, Sydney Medical School, University of Sydney, Sydney, NSW, Australia
e-mail: michelt@chw.edu.au

K.T. Devine
Department of Genetic Medicine, Westmead Hospital, Westmead, NSW, Australia

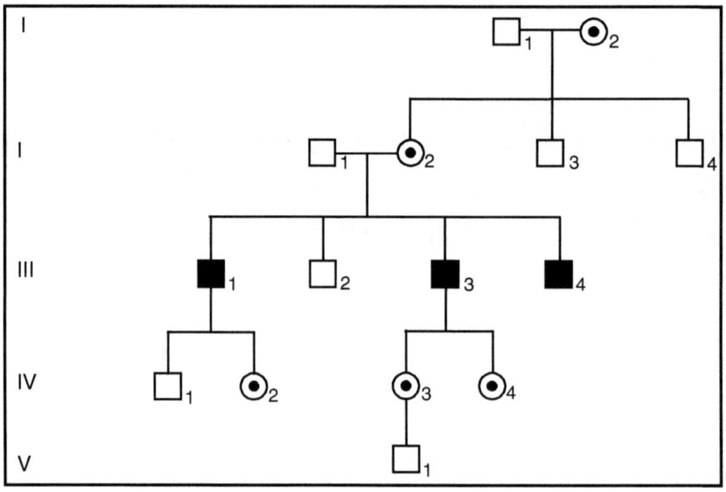

Fig. 1 Family pedigree showing GM (III.1), PM (III.3), and DM (III.4)

in early 2009; however, they all lived at significant distances from our hospital, and thus therapy was instituted at our center initially, but transferred to a local hospital after 4–6 weeks. Many of the monitoring investigations were performed at local institutions, and as a consequence, liver and spleen volumes were not calculated. Treatment compliance was 100% at 12 months.

We report here the clinical phenotype of each patient as well as responses to enzyme replacement therapy (Fig. 4 and Table 1).

Patients

DM (III.4)

DM is a 46-year-old man diagnosed at the age of 20 by the finding of elevated urinary glycospaminoglycans (18 g/mol creatinine), and low enzyme activity levels in plasma (0.05, normal = 4.8 pmol/min/mg protein) and blood leukocytes (1.1, normal = 11.8 pmol/min/mg protein) after he presented with long-standing restriction of joint range of motion. His adult height was 162 cm and he had subtle facial features consistent with a storage disorder (Fig. 3a). DM had joint restriction from childhood; he recalled having difficulty in crossing his legs to sit on the floor in kindergarten and being unable to put his hands on his head. Although he played representative sport during school (including soccer, cricket, and rugby league), he was never a fast runner and always had problems throwing a ball.

DM had acute angle closure glaucoma of the right eye at the age of 20 with intraocular pressure documented at over 80 mmHg. On his most recent ophthalmology review, he had an opaque right cornea and an atrophic optic nerve, with no vision in that eye. His left eye intraocular pressure was maintained at around 14 mmHg with pilocarpine and later latanoprost. He had iridectomy operations performed on both eyes. His left pupil was permanently constricted, presumably from long-standing glaucoma therapy and/or scarring from laser iridectomy, although the effect of GAG storage in the iris sphincter and dilator pupillae muscles may have been contributory. His left eye corrected to 6/7.5 but had severely constricted visual fields. He had no residual night vision and was considered legally blind. Pigmentary retinopathy has not been documented; however, fundoscopy and ERG were not possible due to his right corneal scarring and left miosis.

He presented with parasthesias on neck extension at the age of 30, and MRI demonstrated central spinal canal stenosis in the upper cervical region with abnormal signal intensity at the C2/3 level. The predominant abnormality was described as a greater than 2 mm thickening of a membrane surrounding the dural sac. He underwent surgical decompression at 39 years of age. He underwent left carpal tunnel release at the age of 41 and still has electrophysiological evidence of mild bilateral carpal tunnel syndrome without diminished hand function. DM did not have a history of cardiac symptoms, although echocardiography at the age of 44 showed mild mitral regurgitation and mild aortic sclerosis. The left ventricular and systolic functions were normal and there was mild left ventricular diastolic dysfunction. Respiratory function tests showed a mild obstructive picture [FVC 3.72L (99%), FEV_1 2.3L (74%)], and he had normal exercise tolerance without shortness of breath. DM had mild bilateral high frequency sensorineural hearing loss but did not wear hearing aids.

His other past medical history included Hashimoto's thyroiditis diagnosed at the age of 34, Poland anomaly of the right pectoralis major, a right inguinal hernia repair, and a tonsillectomy. He had a diagnosis of acne rosacea for which he was treated with minocycline. Minocycline caused a

Fig. 2 (**a**) DM (aged 6 months, second from left), PM (aged 22 months, first on left), GM (aged 6 years, third from left), and unaffected brother (aged 4 years, far right). GM's hand contractures were clearly present at that age. (**b**) DM (age 3.5, second from left), GM (age 9 years, third from left), PM (age 5 years, far right) and unaffected brother (age 7 years, far left). Hand contractures are evident in all three affected brothers. (**c**) Unaffected brother (age 20, far left), PM (age 18, second from left), DM (age 16, third from left) and GM (age 22, far right).

striking blue discoloration of his nails and ears – a known complication of this medication (Eisen and Hakim 1998) – and was ceased without resolution of the discoloration.

DM commenced infusion with idursulfase 30 mg weekly at the age of 44. His baseline measurements and changes during the first 12 months of therapy are documented in Fig. 4 and Table 1. There were no infusion reactions or adverse events. DM reported feeling generally fitter and that his joints felt looser, although there were no objective significant changes in his joint range of motion. Liver and spleen volumes were unchanged according to the reports of abdominal CT scans, although organ volumes were not formally quantitated. 6MWT was increased at 12 months compared to baseline (802 m c.f. 564 m). Repeat echocardiography and respiratory function testing after 12 months of idursulfase did not reveal significant changes.

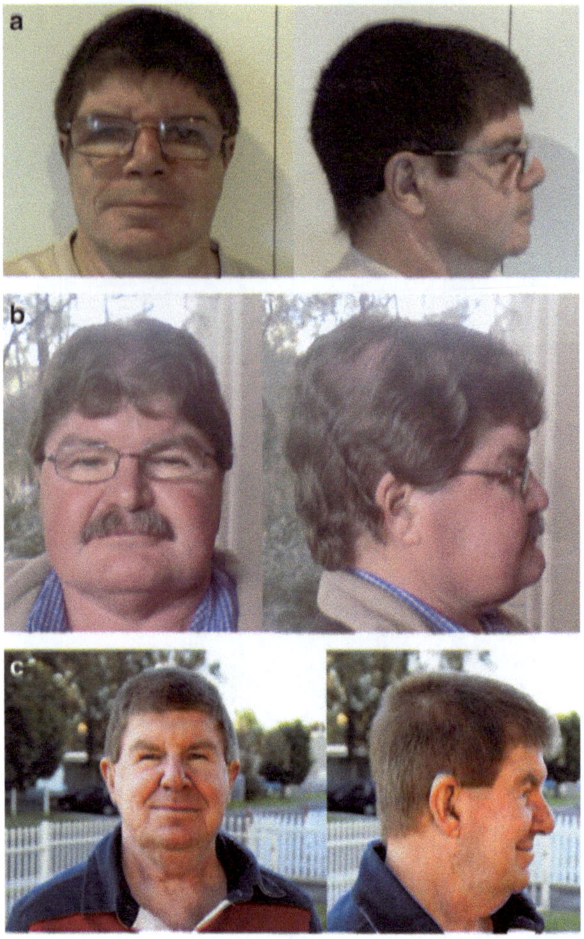

Fig. 3 (**a**) Facial features of DM at the age of 46. (**b**) Facial features of PM at the age of 48. (**c**) Facial features of GM at the age of 52

GM (III.1)

GM is a 52-year-old man diagnosed with Hunter syndrome at the age of 26 following the diagnosis being made in his sibling. His adult height was 160 cm and he also had subtle facial features consistent with his storage disorder (Fig. 3c). GM recalled being unable to cross his arms or legs at the commencement of primary school due to joint restriction.

Cardiac valvular disease predated his diagnosis of MPS II with asymptomatic mitral and aortic valve regurgitation detected as a teenager. Cardiac monitoring was sporadic over the following years, and he presented at the age of 50 in extremis with cardiogenic shock and acute pulmonary edema secondary to severe aortic valve stenosis and mitral regurgitation. He underwent aortic valve replacement with a porcine aortic valve and mitral valve annuloplasty. His postoperative course was complicated by pneumonia, pancreatitis, cardiac tamponade, and *Pseudomonas* sepsis; however, he eventually made a full recovery.

A cervical spine MRI performed at the age of 35 demonstrated dural and extradural thickening from the upper clivus to the C3/4 intervertebral disk level with cord compression and increased signal at the C2 level. Cervical laminectomy was performed at the age of 46 following progression of his chronic neck pain to gait disturbance and bilateral upper limb weakness. Left and right carpal tunnel surgeries were performed at the ages 43 and 47. Baseline respiratory function testing before enzyme replacement therapy at the age of 51 demonstrated mild obstructive and restrictive airways disease [FVC 2.83L (77% predicted), FEV_1 2.10L (70% predicted)]. He had bilateral hearing loss and had worn hearing aids from the age of 43. His other past medical history included tonsillectomy, bilateral inguinal hernia repairs, umbilical hernia, Hashimoto's thyroiditis, and depression.

GM commenced infusion with idursulfase 36 mg weekly at the age of 51. His baseline measurements and changes during the first 12 months of therapy are documented in Fig. 4 and Table 1. There were no infusion reactions or adverse events. He reported an increase in energy levels and exercise tolerance. PM had a significant weight gain of around 10 kg over the first 12 months of therapy. Examination excluded fluid overload from cardiac failure as the cause; abdominal adiposity was felt to be the predominant cause of weight gain. Ophthalmoplegia with restriction of medial and upward gaze bilaterally was noted after 12 months of therapy. Liver and spleen volumes were unchanged according to the reports of abdominal CT scans, although organ volumes were not formally quantitated. 6MWT was 401 m at baseline, 520 m at 6 months, and 425 m after 12 months. Repeat echocardiography at 9 months documented severe dilated cardiomyopathy with good valvular function.

PM (III.3)

PM is a 48-year-old man diagnosed clinically, but not biochemically or molecularly, at the age of 22 following the diagnosis being made in his siblings. The diagnosis was confirmed many years later when his daughters were shown to carry the family mutation. His adult height was 173 cm and he also had subtle facial features consistent with his storage disorder, although these were somewhat masked by a cushingoid appearance from chronic steroid usage (Fig. 3b). PM recalled joint restriction, and a prominent abdomen being present from early childhood. He was never able to pursue physical activity as well as his peers during adolescence and presented with obstructive airways disease at the age of 23.

PM had progressive airways disease and when seen at the age of 46 he had an FVC 2.27L (51%) and an FEV_1 0.72L (20%) corresponding to stage IV COPD (GOLD 2009).

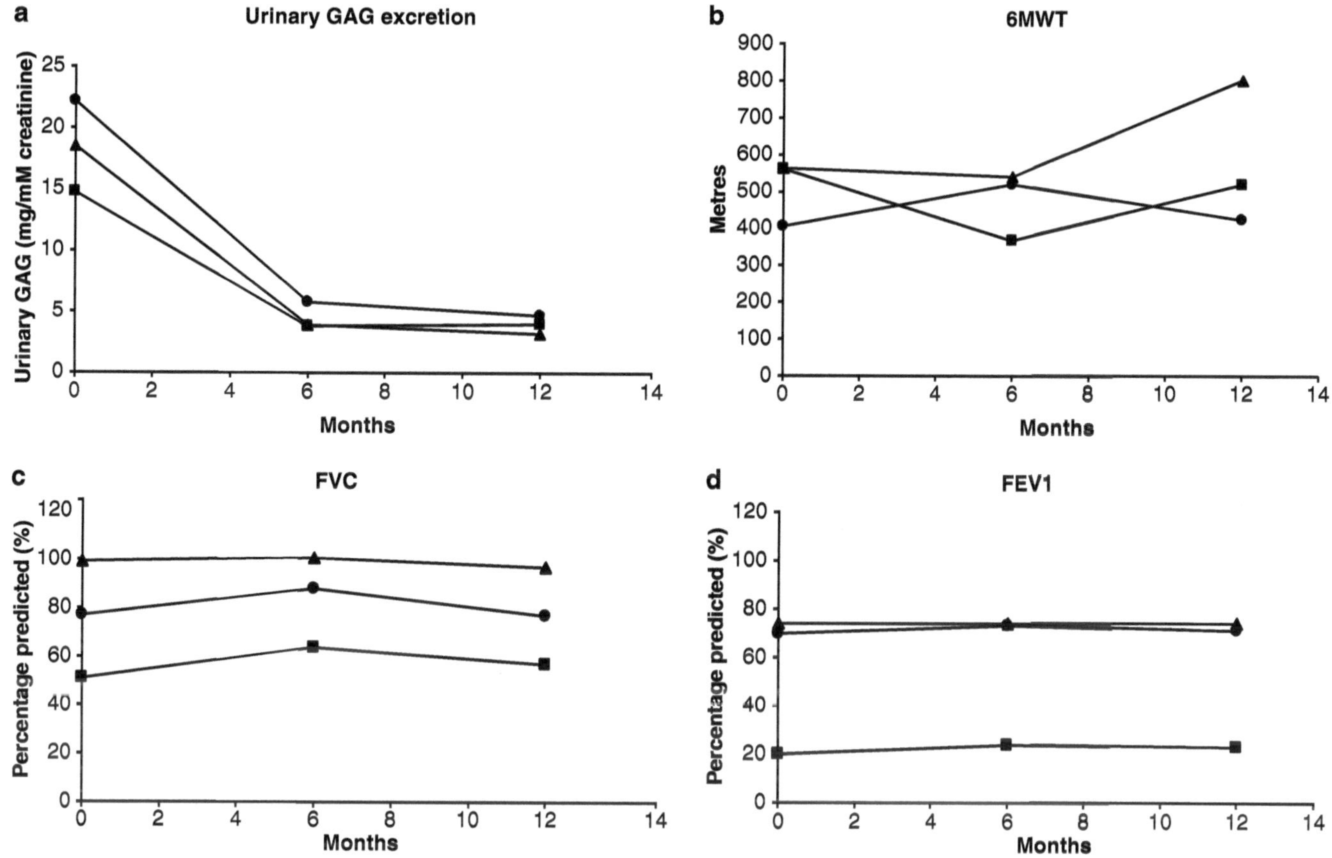

Fig. 4 PM (*square*), GM (*circle*), and DM (*triangle*). (**a**) Urinary GAG excretion; upper limit of normal=8.3 mg/mM creatinine. (**b**) 6MWT. (**c**) FVC. (**d**) FEV_1

He had previously been assessed for combined heart–lung transplantation at the age of 33 but was not considered fit for operation. PM was steroid dependant by the age of 45 and was currently maintained on prednisone 10 mg daily. He had a number of hospitalizations for exacerbations of airways disease, and his exercise tolerance was limited to a few hundred meters on flat ground or ten stairs. He required two rest periods to complete dressing himself each morning. Obstructive sleep apnea was diagnosed at the age of 48 and managed with nocturnal CPAP. High-resolution chest CT demonstrated widespread subtle heterogeneous parenchymal densities with scattered bronchial thickening but no bronchial dilatation. Bands of linear fibrosis were seen in the lower lobes and there was air-trapping present, supportive of the diagnosis of small airways disease. There was no evidence of the paraseptal emphysematous changes we have seen in our adult patients with MPS I (unpublished results). He was a nonsmoker and α_1-antitrypsin deficiency had been excluded.

PM had complete vertical ophthalmoplegia and severe concentric restriction of visual fields. He did not have increased intraocular pressure. New retinal pigmentation was noted at his ophthalmology review 12 months after commencing idursulfase. ERG was not performed. Cervical laminectomy was performed at the age of 43 after he presented with pain, parasthesias, and limb weakness on neck extension. MRI at that time demonstrated spinal canal stenosis from C2 to C4 with diffusely increased cord signal secondary to thickening of the dural and extradural tissues. Surgery did not improve these symptoms, although there was no subsequent progression of his pain, parasthesias, or weakness. Bilateral carpal tunnel surgery was performed at the age of 22. PM had significant hearing loss and was deaf in the left ear and wore a hearing aid in the right ear. Echocardiography demonstrated aortic valve stenosis (valve area 1.2 cm^2) with normal systolic function and mild diastolic dysfunction.

PM commenced weekly 36 mg idursulfase at the age of 46. His baseline measurements and changes during the first 12 months of therapy are documented in Fig. 4 and Table 1. There were no infusion reactions or adverse events. He traveled an hour each way to his treatment center each

Table 1 Joint range of motion measurements made at baseline and 12 months. Each joint was measured bilaterally and the two results averaged (except for flexion and extension of the neck). Results are presented as the mean degrees ± standard error of the mean. The *p* values were calculated using a two-tailed Student's *t*-test. There was no significant change in joint range of motion following 12 months of idursulfase

	Baseline					12 months					Change	*p* Value
	GM	PM	DM	Mean	SE	GM	PM	DM	Mean	SE		
Ankle												
Dorsiflexion	2.5	−12.5	0.0	−3.3	±4.6	4.5	−17.5	5.5	−2.5	±7.5	0.8	0.8125
Plantarflexion	49.5	52.5	43.5	48.5	±2.6	46.0	40.0	45.5	43.8	±1.9	4.7	0.3846
Inversion	4.0	5.5	5.0	4.8	±0.4	8.0	0.0	19.0	9.0	±5.5	4.2	0.5363
Eversion	11.0	24.5	19.5	18.3	±3.9	12.5	0.0	24.5	12.3	±7.1	6.0	0.5851
Knee												
Flexion	122.5	127.5	138.5	129.5	±4.7	122.0	125.0	127.0	124.7	±1.5	4.8	0.2893
Extension	−11.0	2.5	−3.5	−4.0	±3.9	−10.5	0.0	−4.0	−4.8	±3.1	0.8	0.4444
Hip												
Flexion	86.5	100.0	101.0	95.8	±4.7	96.0	110.0	96.0	100.7	±4.7	4.8	0.4294
Extension	−9.5	0.0	−2.0	−3.8	±2.9	−11.0	1.0	−12.0	−7.3	±4.2	3.5	0.4034
Abduction	24.5	20.0	20.5	21.7	±1.4	31.0	20.0	37.5	29.5	±5.1	7.8	0.2545
Adduction	13.0	12.5	21.0	15.5	±2.8	22.0	17.5	19.5	19.7	±1.3	4.2	0.3064
Internal rotation	0.5	−2.0	11.0	3.2	±4.0	0.0	2.5	8.0	3.5	±2.4	0.3	0.8937
External rotation	30.5	35.0	22.0	29.2	±3.8	35.0	27.5	15.0	25.8	±5.8	3.3	0.4846
Wrist												
Flexion	48.0	54.0	76.0	59.3	±8.5	67.5	52.5	46.5	55.5	±6.2	3.8	0.8124
Extension	42.0	51.0	41.0	44.7	±3.2	51.0	40.0	44.0	45.0	±3.2	0.3	0.9603
Ulnar deviation	39.5	36.5	32.5	36.2	±2.0	37.5	27.0	36.5	33.7	±3.3	2.5	0.5876
Radial deviation	11.5	1.5	10.5	7.8	±3.2	12.5	7.0	18.5	12.7	±3.3	4.8	0.1422
Elbow												
Flexion	140.0	137.0	145.0	140.7	±2.3	140.0	134.0	147.5	140.5	±3.9	0.2	0.9261
Extension	−15.0	−35.0	−24.0	−24.7	±5.8	−17.5	−26.0	−26.5	−23.3	±2.9	1.3	0.7612
Shoulder												
Flexion	120.0	116.0	110.0	115.3	±2.9	115.0	105.0	135.0	118.3	±8.8	3.0	0.8129
Extension	58.0	50.0	65.0	57.7	±4.3	43.5	40.0	61.0	48.2	±6.5	9.5	0.0890
Abduction	92.5	69.5	124.0	95.3	±15.8	85.0	98.0	125.0	102.7	±11.8	7.3	0.5692
Internal rotation	26.0	15.5	23.5	21.7	±3.2	55.0	20.0	21.0	32.0	±11.5	10.3	0.3923
External rotation	19.0	54.0	72.5	48.5	±15.7	0.0	72.5	70.0	47.5	±23.8	1.0	0.9350
Neck												
Flexion	39.0	40.0	45.0	41.3	±1.9	45.0	40.0	35.0	40.0	±2.9	1.3	0.8020
Extension	23.0	26.0	54.0	34.3	±9.9	35.0	30.0	70.0	45.0	±12.6	10.7	0.0942
Rotation	41.0	60.0	39.0	46.7	±6.7	37.5	50.0	61.5	49.7	±6.9	3.0	0.7911
Lateral flexion	20.0	26.0	18.5	21.5	±2.3	16.5	20.0	23.5	20.0	±2.0	1.5	0.6964

week and felt exhausted following each infusion, such that he would sleep the rest of the day and be unable to perform general tasks around the home the next day. He felt that treatment had not increased his exercise tolerance and that it had a significantly detrimental effect on his quality of life. He ceased treatment following his 12-month review.

Discussion

These three brothers demonstrate the attenuated spectrum of disease associated with MPS II and the variability that may be present. Each has had similar manifestations with regard to the organ systems affected, but there has been striking variability in the severity of each disease symptom. A response to enzyme replacement therapy was not demonstrable on objective measurements, although for GM and DM there were subjective improvements in energy and exercise tolerance. For PM, the burden of weekly therapy was more than any perceived benefit.

All three have had serious skeletal and connective tissue manifestations of disease including joint restriction, cervical spine stenosis, and carpal tunnel syndrome. These have all been described as common in published series of attenuated MPS II patients (Young and Harper 1982; Kulkarni et al. 1987; Parsons et al. 1996; Wraith et al. 2008a, b). In contrast to descriptions in the literature in which hip dysplasia is a common complaint, it was not present clinically in these patients, although radiographs were not performed to exclude subclinical involvement.

All three brothers had ocular manifestations. DM presented before diagnosis with an episode of acute angle closure glaucoma that rendered him blind in his right eye.

Despite good subsequent control of intraocular pressure, he had constricted visual fields and night blindness in the contralateral eye. Both PM and GM had partial ophthalmoplegia, and PM also had retinal pigmentation and severely restricted visual fields. Other ocular changes in MPS II, not seen in these brothers, include exophthalmos and disk swelling with optic atrophy (Ashworth et al. 2006a, b).

We believe this is the first description of ophthalmoplegia in MPS II, although a single patient with MPS I has been described with mechanical limitation of superior oblique tendon movement (Bradbury et al. 1989). In that case, it was thought to be due to tendon shortening, and this could also be the mechanism in our patients, although deposition of storage material producing a mass effect in and on the extraocular muscles is also possible.

Ocular hypertension and glaucoma have been documented in the mucopolysaccharidoses and are due to deposition of storage material in the anterior chamber structures and obstruction of outflow through the trabecular network (McDonnell et al. 1985; Ashworth et al. 2006a, b). Chronic closed angle, acute closed angle, and open angle glaucoma have all been documented, although there is only one previously described case of acute closed angle glaucoma in MPS II (Kaiden 1982). A review of 50 patients with MPS documented raised intraocular pressure in nine (seven of whom had MPS VI), although none had acute glaucoma as our patient did (Ashworth et al. 2006a, b).

Pigmentary retinopathy is well documented in the MPS disorders, particularly in MPS III (Ashworth et al. 2006a, b), and in advanced cases causes night blindness and loss of peripheral vision. Five of ten patients in the JET study had retinal degeneration (Okuyama et al. 2009). Although retinal pigmentation had not been documented in PM before the completion of 12 months therapy with idursulfase, the history of restricted visual fields argues that he had pigmentary retinopathy of a long-standing nature. DM also had restricted visual fields and night blindness in the left eye, both suggestive of a pigmentary retinopathy although this has not been demonstrated.

Cardiac valvular disease is a nearly ubiquitous feature in older patients (Young and Harper 1982); even in the younger cohort of the Hunter Outcome Survey, it was present in 57% of patients and manifested at a median age of 6.1 years (Wraith et al. 2008a, b). The variability in severity of this manifestation between these brothers was striking, with GM manifesting an almost fatal degree of aortic stenosis and DM having minimal disease.

PM had severe obstructive airways disease with newly diagnosed obstructive sleep apnea. This is in contrast to the commonly described airways disease of MPS II which is an upper airways crowding secondary to soft tissue enlargement, followed by a restrictive pattern due to thoracic skeletal involvement (Wraith et al. 2008a, b). PM's lung disease was very severe, and it should be noted that both DM and GM also had mild airways disease.

It is interesting that both GM and DM had Hashimoto's thyroiditis; however, we do not feel that this is likely to represent an association with MPS II, as autoimmune thyroid disease is a common disorder with familial clustering and a complex genetic background, with recent studies implicating numerous susceptibility loci (Tomer 2010).

There was a lack of objectively measurable clinical response to idursulfase in these brothers, although there were clear biochemical responses with normalization of urinary GAG's in the first 6 months. Comparisons between the response to idursulfase in these siblings and those of the clinical trial patients are difficult to make, particularly because of the significant age difference. The oldest phase I/II/III trial patient was 31 years old, and around half were prepubertal at the time of baseline investigations (Muenzer et al. 2006; 2007). A more meaningful comparison can be made with the patients in the JET study; an investigation of ten Japanese adults with MPS II following 12 months therapy with idursulfase in whom the mean age was 31 and the oldest patient was 53 (Okuyama et al. 2009). Patients in the JET study significantly improved their liver and spleen volumes, normalized urinary GAG's, and improved 6MWT and FVC although these last two measurements were not increased to statistically significant levels. Similarly, these brothers normalized their urinary GAG excretion but did not have significant changes in respiratory or cardiac function. The variability in results of the 6MWT demonstrates that patient effort plays a significant role in the reliability and reproducibility of this test, and that as a measure of clinical efficacy its usefulness is limited in individual patients. It is likely that the age of the patients and the correspondingly long-standing nature of their disease made it difficult to reverse measurable symptoms over this initial 12-month period of therapy.

Conclusion

MPS II in these adult siblings has caused significant morbidity, albeit in different ways for each patient: GM had near fatal cardiac valvular dysfunction, PM has severe chronic obstructive airways disease, and DM is legally blind. These siblings demonstrate the multisystem nature of this disorder and its variable effects even within members of the same family.

GM and DM felt they had increased energy and exercise tolerance following 12 months of treatment with idursulfase. In contrast, PM did not feel improved on therapy and felt that the weekly burden of traveling a long distance to receive therapy and the associated fatigue resulted in a decrease in his quality of life. There were no infusion reactions or adverse events, and over 12 months of therapy the clinical status of the patients remained stable.

In adults with long-standing accumulation of GAG's and a resulting significant burden of disease, it is likely that objectively measurable improvements in clinical outcomes will be slow to achieve; however, achieving clinical stability in these patients is a lifesaving outcome. The importance of early diagnosis, which could have been made at a young age in this family, should be emphasized as enzyme replacement therapy is more likely to be effective in preventing, rather than reversing, the serious sequelae of this multisystem disease.

Acknowledgments We thank the three brothers for their cooperation with this study. We acknowledge Professor Gillian Turner who made the diagnosis in this family.

Synopsis

Three adult siblings with MPS II demonstrated the striking variability of this disorder; we documented the findings after 12 months of enzyme replacement therapy.

Author Contributions

Michel Tchan wrote the initial draft article and performed examinations and investigations on the patients. Guarantor.

Kerry Devine administered enzyme replacement therapy and provided revision and criticism of the article.

David Sillence revised the draft article and supervised enzyme replacement therapy.

Competing Interest Statement

Dr. Tchan reports having received reimbursement from Genzyme Corporation for travel expenses to attend educational meetings.

Details of Funding

No funding was obtained for this study.

Ethics Approval

No ethics approval was required for this study.

Patient Consent

Consent was obtained from the family, including consent for publication of photographs.

References

Ashworth JL, Biswas S, Wraith E et al (2006a) Mucopolysaccharidoses and the eye. Surv Ophthalmol 51(1):1–17

Ashworth JL, Biswas S, Wraith E et al (2006b) The ocular features of the mucopolysaccharidoses. Eye (Lond) 20(5):553–563

Baehner F, Schmiedeskamp C, Krummenauer F et al (2005) Cumulative incidence rates of the mucopolysaccharidoses in Germany. J Inherit Metab Dis 28(6):1011–1017

Bradbury JA, Martin L, Strachan IM (1989) Acquired Brown's syndrome associated with Hurler-Scheie's syndrome. Br J Ophthalmol 73(4):305–308

Eisen D, Hakim MD (1998) Minocycline-induced pigmentation. Incidence, prevention and management. Drug Saf 18(6):431–440

GOLD (2009) Global strategy for the diagnosis, management and prevention of chronic obstructive pulmonary disease. The GOLD report

Kaiden JS (1982) Angle closure in a patient with Hunter's syndrome. J Ocul Ther Surg 1:250

Kulkarni MV, Williams JC, Yeakley JW et al (1987) Magnetic resonance imaging in the diagnosis of the cranio-cervical manifestations of the mucopolysaccharidoses. Magn Reson Imaging 5(5):317–323

Manara R, Rampazzo A, Cananzi M et al (2010) Hunter syndrome in an 11-year old girl on enzyme replacement therapy with idursulfase: brain magnetic resonance imaging features and evolution. J Inherit Metab Dis Jan 6 (online)

McDonnell JM, Green WR, Maumenee IH (1985) Ocular histopathology of systemic mucopolysaccharidosis, type II-A (Hunter syndrome, severe). Ophthalmology 92(12):1772–1779

Muenzer J, Wraith JE, Beck M et al (2006) A phase II/III clinical study of enzyme replacement therapy with idursulfase in mucopolysaccharidosis II (Hunter syndrome). Genet Med 8(8):465–473

Muenzer J, Gucsavas-Calikoglu M, McCandless SE et al (2007) A phase I/II clinical trial of enzyme replacement therapy in mucopolysaccharidosis II (Hunter syndrome). Mol Genet Metab 90(3):329–337

Okuyama T, Tanaka A, Suzuki Y et al (2009) Japan Elaprase treatment (JET) study: idursulfase enzyme replacement therapy in adult patients with attenuated Hunter syndrome (mucopolysaccharidosis II, MPS II). Mol Genet Metab 99(1):18–25

Parsons VJ, Hughes DG, Wraith JE (1996) Magnetic resonance imaging of the brain, neck and cervical spine in mild Hunter's syndrome (mucopolysaccharidoses type II). Clin Radiol 51(10):719–723

Poorthuis BJ, Wevers RA, Kleijer WJ et al (1999) The frequency of lysosomal storage diseases in The Netherlands. Hum Genet 105(1–2):151–156

Tomer Y (2010) Genetic susceptibility to autoimmune thyroid disease: past, present, and future. Thyroid 20(7):715–725

Wraith JE, Beck M, Giugliani R et al (2008a) Initial report from the Hunter outcome survey. Genet Med 10(7):508–516

Wraith JE, Scarpa M, Beck M et al (2008b) Mucopolysaccharidosis type II (Hunter syndrome): a clinical review and recommendations for treatment in the era of enzyme replacement therapy. Eur J Pediatr 167(3):267–277

Young ID, Harper PS (1982) Mild form of Hunter's syndrome: clinical delineation based on 31 cases. Arch Dis Child 57(11):828–836

JIMD Reports
DOI: 10.1007/8904_2011_18

Received: 15 December 2010; Revised: 7 February 2011;
Accepted: 21 February 2011; Published online: 22 June
Communicated by: Eva Morava

Two Argentinean Siblings with CDG-Ix: A Novel Type of Congenital Disorder of Glycosylation?

M.B. Bistué Millón, M.A. Delgado, N.B. Azar, N. Guelbert, L. Sturiale, D. Garozzo, G. Matthijs, J. Jaeken, Raquel Dodelson de Kremer, and C.G. Asteggiano

Abstract Congenital disorders of glycosylation (CDG) are genetic diseases caused by abnormal protein and lipid glycosylation. In this chapter, we report the clinical, biochemical, and molecular findings in two siblings with an unidentified CDG (CDG-Ix). They are the first and the third child of healthy consanguineous Argentinean parents. Patient 1 is now a 11-year-old girl, and patient 2 died at the age of 4 months. Their clinical picture involved liver dysfunction in the neonatal period, psychomotor retardation, microcephaly, seizures, axial hypotonia, feeding difficulties, and hepatomegaly. Patient 1 also developed strabismus and cataract. They showed a type 1 pattern of serum sialotransferrin. Enzymatic analysis for phosphomannomutase and phosphomannose isomerase in leukocytes and fibroblasts excluded PMM2-CDG and MPI-CDG. Lipid-linked oligosaccharide (LLO) analysis showed a normal profile. Therefore, this result could point to a deficiency in the dolichol metabolism. In this context, *ALG8-CDG*, *DPAGT1-CDG*, and *SRD5A3-CDG* were analyzed and no defects were identified. In conclusion, we could not identify the genetic deficiency in these patients yet. Further studies are underway to identify the basic defect in them, taking into account the new CDG types that have been recently described.

Keywords Congenital disorders of glycosylation · Isoelectrofocusing · N- and O-glycosylation

M.B.B. Millón, M.A. Delgado, N.B. Azar, N. Guelbert, and R.D. de Kremer
Centro de Estudio Metabolopatías Congénitas (CEMECO), Universidad Nacional de Córdoba, Hospital de Niños de la Santísima Trinidad, Ferroviarios 1250, CP X5014AKN Cordoba, Argentina

L. Sturiale and D. Garozzo
Istituto di Chimica e Tecnologia dei Polimeri, CNR, Catania, Italy

G. Matthijs
Center for Human Genetics, Katholieke Universiteit Leuven, Leuven, Belgium

J. Jaeken
Center for Metabolic Diseases, Katholieke Universiteit Leuven, Leuven, Belgium

C.G. Asteggiano (✉)
Centro de Estudio Metabolopatías Congénitas (CEMECO), Universidad Nacional de Córdoba, Hospital de Niños de la Santísima Trinidad, Ferroviarios 1250, CP X5014AKN Cordoba, Argentina
and
Cátedra de Química Biológica, Universidad Católica de Córdoba, Córdoba, Argentina
and
Consejo Nacional de Investigaciones Científicas y Técnicas (CONICET), Buenos Aires, Argentina
e-mail: casteggi@campus1.uccor.edu.ar; asteggianocarla@hotmail.com

Abbreviation

CDG	Congenital disorder of glycosylation
COG	Oligomeric golgi complex
HPLC	High-performance liquid chromatography
IEF	Isoelectrofocusing
MS	Mass spectrometry
Tf	Transferrin

Introduction

Congenital disorders of glycosylation (CDG) are genetic defects in the synthesis and transfer of the glycan moiety of glycoproteins and glycolipids. About ~1% of the human genome is estimated to be involved in glycosylation processes (Jaeken 2003; Morava et al. 2008). Disorders of N-glycosylation are divided into CDG type I or II, according to the intracellular localization of the molecular defect (McKenzie et al. 2007; Morava et al. 2008; Lefeber et al. 2009; Jaeken et al. 2009). CDG-I is caused by defects in enzymes governing the synthesis and transfer of the oligosaccharide in the ER. On the other hand, defects leading to CDG-II belong to different classes: enzymes responsible for the modifications of the *N*-glycan chain in the Golgi apparatus, sugar transporters, and proteins with a role in intracellular trafficking, like the COG complex. A new nomenclature system indicates the specific disorder with the use of the gene name or the protein name to include all protein glycosylation disorders and to extend it to the lipid glycosylation defects (Jaeken et al. 2008). The diseases are labeled CDG-Ix as long as the basic defect remains unknown.

Since the initial description of CDG in 1980 (Jaeken et al. 1980), serum transferrin IEF with immunodetection has been widely used as a screening test. Any defect in the synthesis or processing of these glycans results in the alteration of sialotransferrin isoforms, which is detectable by IEF according to their different charges (Jaeken 2003). The analysis on HPLC allows for the separation of transferrin glycoforms based not only on the net charge of the molecule, but also on structural differences of the glycans. Both methodologies can be used for assignment of cases to either type I or type II (Sturiale et al. 2008; Quintana et al. 2009).

Application of matrix-assisted laser desorption/ionization time of flight mass spectrometry (MALDI-TOF MS) allows the high-throughput identification of proteins by the accurate mass measurement of peptides derived from total proteome digestion (Reinders et al. 2004; Guerrera and Kleiner 2005). MALDI-TOF applications in CDG aim to investigate glycosylation changes of proteins, providing structural information on the targeted protein. In addition, MS characterization of defective glycan structures is an essential step to solve defects in patients with CDG-x (Sturiale et al. 2005; Babovic-Vuksanovic and O'Brien 2007). The clinical spectrum of N-glycosylation defects is variable, ranging from severe multisystem disorders to dysfunction of specific organs. The clinical, metabolic, and molecular aspects of CDG patients have been described extensively in recent reviews (Marklová and Albahri 2007; de Lonlay et al. 2008; Barone et al. 2009). Other reports (Marklová and Albahri 2007; Morava et al. 2008) tried to find similarities in the clinical presentations of patients and to detect specific symptoms suggestive of a distinct CDG I type. Clinical features such as inverted nipples and abnormal fat distribution may be found in PMM2-CDG (CDG Ia) (OMIM 212065) patients, the most frequent type. Other clinical features such as retrognathia, low-set ears, and club foot are common findings in congenital central nervous system anomalies. In one of the studies, an overview of the unusual clinical symptoms in CDG-Ix (OMIM 212067) showed that two main subgroups could be distinguished based on the severity of the disease: one with a pure neurological presentation and the other with a neurological-multivisceral form (Morava et al. 2008). Nevertheless, other unique findings in CDG-Ix include arthrogryposis, macrocephaly, polyneuropathy, and cystic kidneys.

We want to report the clinical features, the biochemical studies, and the molecular analysis of two siblings characterized as CDG-Ix, who have been identified in the context of the first program for these pathologies in Argentina.

Case Report

We present two sisters born as the first and the third child of healthy consanguineous Argentinean parents. One of the children is now 11 years old (patient 1) and the other (patient 2) had an early death at the age of 4 months. The clinical features of both patients, who share some common features of CDG, are illustrated in Fig. 1a, b. The clinical findings include axial hypotonia, psychomotor retardation, dysmorphic features, feeding problems, hepatomegaly, coagulopathy, and recurrent infections (Table 1). Patient 2 was born with a severe phenotype consisting of liver involvement (hepatomegaly, hypoalbuminemia, protein-losing enteropathy, recurrent vomiting, and diarrhea), seizures, progressive developmental delay, and coagulation abnormalities with recurrent infections. Unfortunately, detailed examination of the brain was not possible due to the early death (Table 1). In patient 1, ophthalmological abnormalities were observed including strabismus and nuclear-cortical cataracts (Fig. 1a). She presented liver dysfunction in the neonatal period, and has had psychomotor retardation, microcephaly, intractable seizures, feeding difficulties, and malnutrition.

Materials and Methods

Blood samples from both patients were obtained after informed parental consent. EDTA blood samples were taken from patient 1 and the parents to extract genomic DNA from leukocytes with a commercially available kit (Wizard Genomic Purification Kit, Promega, Madison, WI). For cell culture, skin biopsy was obtained from patient 1, and the fibroblasts were cultivated in minimal essential medium (MEM). Approval of Human Research was obtained from the institutional review boards of CIEIS-Health Investigation Ethic Committee. Children's Hospital, Córdoba, Argentina.

SDS-PAGE and Western Blot

Plasma sample from P1 was incubated for 30 min at room temperature with a solution of NaCl 0.9% and 10 mM ferric citrate in a ratio of 15:70:15 to saturate the transferrin with iron. The saturated sample was diluted 1:20 in sample

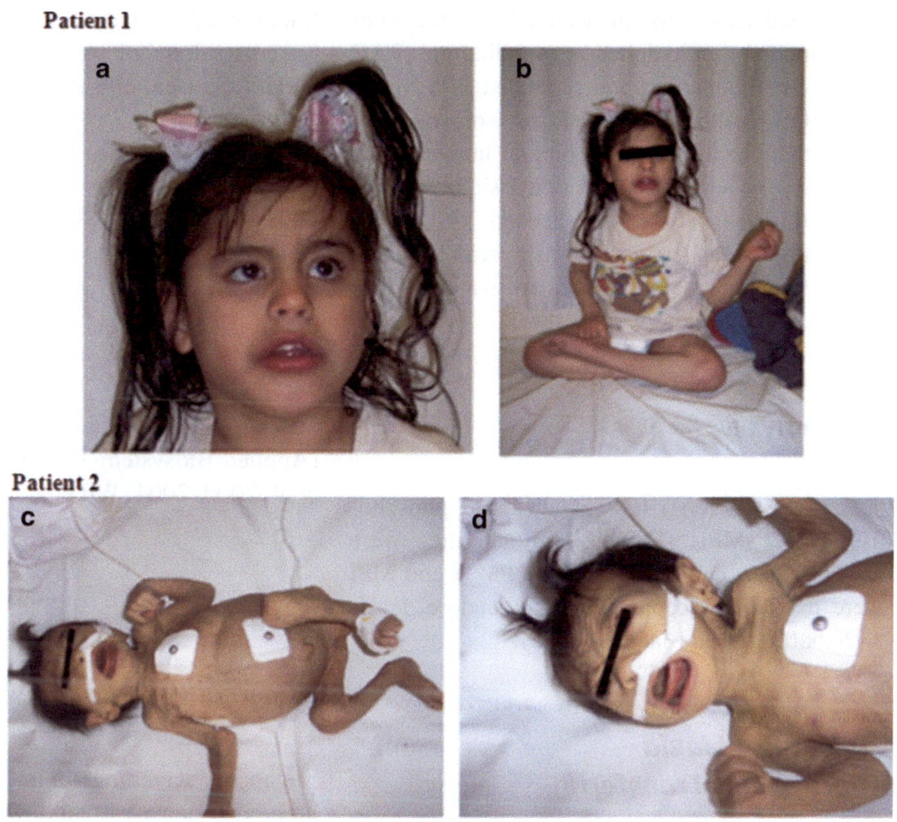

Fig. 1 Clinical features of two CDG-Ix sibling patients. Patient 1 (**a** and **b**) at 8 years old. She showed psychomotor retardation, feeding problems, hepatomegaly and ophthalmological abnormalities (strabismus and cataract). Patient 2 (**c** and **d**): she had dysmorphic features and a severe phenotype consisting of liver involvement (hepatomegaly, hypoalbuminemia, ascitis, protein-losing enteropathy, recurrent vomiting, and diarrhea), microcephaly, intractable seizures, axial hypotonia, feeding difficulties, and malnutrition. She had recurrent infections and an early death at 4 months

Table 1 Overview of the clinical symptoms observed in two siblings CDG-Ix

Clinical symptoms	Patient 1 (11 years old)	Patient 2 early death (at 4 months)
Psychomotor retardation	+++	+++
Cerebellar atrophy	–	nd
Seizures	+++ (refractory)	+
Stroke like episodes	–	–
Extrapyramidal symptoms	+	–
Cataract	+ (nuclear-cortical, and bilateral)	+
Strabismus	+	–
Optic atrophy	–	–
Liver involvement	+	+++
Failure to thrive	++	+
Diarrhea/vomiting	–	++
Hypoalbuminemia	–	++
Cardiomyopathy	–	–
Cystic kidney disease	–	nd
Hypotonia	Axial	Axial
Muscle weakness, CK levels	+ CK: normal	++ CK: nd
Coagulation anomalies: prothrombin (PT) and Kaolin partial thromboplastin time (KPTT)	+ PT: 52 seg (NV:11.5–13.5 seg) KPPT: 49.6 seg (VN= 25–34 seg)	+
Thrombocytopenia	–	+
Respiratory failure	–	–
Recurrent infections	+ (urinary and respiratory)	++

++ very severe, + symptom present, – symptom absent, *nd* not determined

treatment buffer containing 250 μl Tris–HCl 1.25 mol/L (pH: 6.8), 400 μl SDS 25%, 250 μl 2-mercaptoethanol, 500 μl glycerol 99%, 250 μl 0.2% bromophenol blue, and 3,350 μl of bidistilled water (final volume 5 ml; pH 6.8) and separated on 10% SDS-PAGE and transferred to a PVDF membrane by western blotting using standard procedures (Artuch et al. 2003). The blocked membrane was incubated with polyclonal antibodies for anti-transferrin and anti-haptoglobin glycoproteins (Dako, Germany) using an anti-rabbit peroxidase-goat conjugated as secondary antibody (Bio Rad). The visualization of the protein bands was performed using a colorimetric detection (DAB, Sigma).

IEF of Serum Transferrin (Tf-IEF)

Plasma sample from P1 (20 μl) was incubated for 30 min at room temperature with a solution of NaCl 0.9% (20 μl), 10 mM ferric citrate (2 μl), and 0.5 mM sodium hydrogen carbonate (2 μl) to saturate the transferrin with iron. After centrifugation ($2{,}000 \times g$, 10 min), the supernatant was further diluted (1:5) with water.

Proteins of P2 were extracted by incubating blood spots samples for 1 h at room temperature with a solution of NaCl 0.9% (95 μl) and 10 mM ferric citrate (20 μl) and centrifugated at $3{,}500 \times g$ for 10 min.

The iron-saturated serum proteins were diluted five times with water and applied to a hydrated immobiline gel (PAG plate pH 4–6.5; GE Healthcare) and separated in a Multiphore II system (GE Healthcare). Transferrin isoforms were detected after immunofixation with rabbit anti-human transferrin antibody (Dako, Germany) and Coomassie blue staining (Jaeken et al. 1984; Stibler and Jaeken 1990). The relative amounts of the transferrin isoforms were determined and quantified using the Image J 1.42q Software (Wayne Rasband National Institutes of Health, USA).

Enzymatic Analysis

Phosphomannomutase (PMM) (EC 5.4.2.8) and phosphomannose isomerase (MPI) (EC 5.3.1.8) activities were measured in leukocytes and fibroblast. The PMM and MPI activities were performed according to a procedure developed by van Schaftingen and Jaeken (1995).

High-Performance Liquid Chromatography of Transferrin

High-performance liquid chromatography of transferrin (Tf-HPLC) analysis was based on the method described by Helander et al. (2003).The HPLC system consisted of an Agilent 1100 Series liquid chromatography. Separation of the transferrin glycoforms was performed on a SOURCE 15Q PE 4.6/100 anion-exchange chromatography column (GE Healthcare) at 25°C by linear salt gradient elution at a flow rate of 1.0 ml/min. Quantification of the transferrin glycoforms relied on the selective absorbance of the iron–transferrin complex at 470 nm. The relative amount of each transferrin isoform was expressed as a percentage of the area under the curve (%AUC) (Helander et al. 2003; Quintana et al. 2009).

Intact transferrin (immunopurified as described in Sturiale et al. 2008) and its relative *N*-glycan pool, released by peptide-*N*-glycosidase F [PNGase F (EC 3.5.1.52)] treatment, were both analyzed by MALDI MS on a Voyager STR instrument (Applied Biosystems, Framingham, MA) using sinapinic acid as matrix for native protein analysis, and 2′, 4′, 6′-trihydroxyacetophenone (THAP) for glycan profiling (Sturiale et al. 2008).

Lipid-Linked Oligosaccharide Analysis

Lipid-linked oligosaccharide (LLO) analysis was measured in fibroblasts after metabolic labeling. They were size-fractionated and analyzed by HPLC. Oligosaccharides linked to dolichol were released, extracted, and analyzed by HPLC essentially as described by Denecke et al. (2005).

Molecular Studies

Genomic DNA was obtained from whole blood or fibroblasts using a commercially available kit (Qiagen, Hilden, Germany). Genomic DNA was amplified using the primers for the *PMM2*, *MPI*, *ALG8*, *DPAGT1*, and *SRD5A3* genes and analyzed by direct sequencing in an AB3130 system (Applied Biosystems) (Matthijs et al. 1997, 1998; Schollen et al. 2000, 2004; Wu et al. 2003; Cantagrel et al. 2010).

Results

Biochemical Studies

Metabolic screening (amino acids, urinary organic acids, ammonia, lactate, blood pH, and very long-chain fatty acids) was normal. We found an increased serum activity of some lysosomal enzymes. Patient 1 has high levels of arylsulphatase A (EC 3.1.6.8) (53.87 nmol/h/ml) (NV: 9–30 nmol/h/ml), and patient 2 showed very high levels of β-hexosaminidase (EC 3.2.1.52) (1686,8nmol/h/ml) (VN: 140–600 nmol/h/ml); however, they had normal levels of β-glucuronidase (EC 3.2.1.31).

During a temporary and transient hepatic problem (altered urine organic acids and plasma amino acids consistent with mild liver dysfunction), we found slightly increased levels of serum galactose, with normal enzymatic activities of galactose-1-phosphate-uridyltransferase [GALT (EC 2.7.7.12)] and galactokinase (EC 2.7.1.6), excluding classic galactosemia and galactokinase deficiency in patient 1. This increase was not observed in subsequent determinations of galactose levels in dry blood samples. Moreover, gas chromatographic analysis of galactitol and galactose in urine was normal.

Transferrin Analysis

Western blot of two serum glycoproteins (transferrin and haptoglobin) showed an altered profile similar to the CDG Type I pattern (Fig. 2b). Furthermore, Tf-IEF also showed a type I transferrin pattern in both patients (Fig. 2a), with increased asialo- and disialotransferrin isoforms and

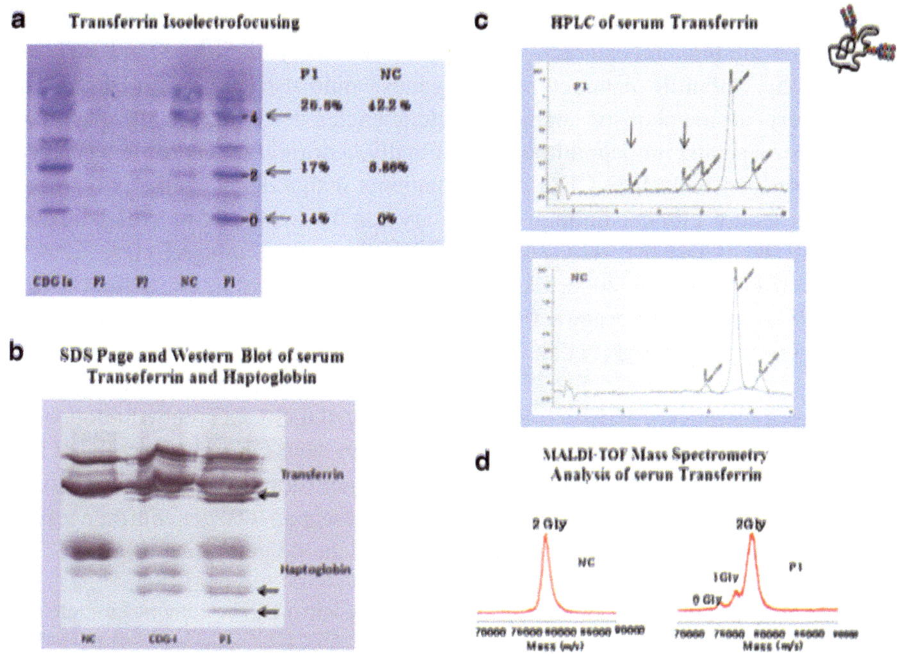

Fig. 2 (**a**) Transferrin IEF pattern of two CDG-Ix patients: *P1* serum of patient 1, *P2* dried blood spot samples of patient 2, *CDG Ia* control of a PMM2-CDG patient, and *NC* a healthy control. The *arrows* on the *right* indicate tetrasialo, disialo, and asialotransferrin, carrying two, one, or no oligosaccharide chains. The comparison of transferrin isoform values (%) from patient 1 with healthy control showed an increase in the cathodic fractions (asialo- and disialo-) and a decrease in tetrasialo- of the patient respect to the healthy control. (**b**) SDS-PAGE profile of serum proteins from patient 1 in comparison with a healthy control (NC) and a CDG type I patient. The *arrows* indicate the increased bands corresponding to abnormal sialotransferrins. (**c**) High-performance liquid chromatography (HPLC) profile of serum transferrin of patient 1 (at the *top*) and of a healthy control (NC at the *bottom*) shows the presence of underglycosylated isoforms in the patient. The *arrows* indicate the increased isoforms corresponding to asialo- and disialotrasferrin. (d) MALDI-TOF mass spectrum of serum transferrin from P1 (on the *right*) and from a healthy control (on the *left*). Patient 1 showed abnormal isoforms corresponding to mono-glycosylated and a-glycosylated species

decreased tetrasialotransferrin (14% asialo-, 17% disialo-, and 25% tetrasialotransferrin; reference values: n/d, 3.8% and 46% for asialo-, disialo-, and tetrasialotransferrin, *n*: 24). IEF of serum Tf after neuraminidase treatment showed only one band excluding a Tf protein variant. The abnormal Tf-IEF pattern was confirmed by HPLC and MALDI-TOF MS in patient 1. She presented a clear increase in asialo- and disialotransferrin peaks: 1.43% and 3.89%, respectively [reference values: disialotransferrin: 0.28% and not detected for asialotransferrin in normal controls (*n*) *n*: 24 (Fig. 2c)]. Tf analysis by MALDI MS showed an abnormal glycosylation profile due to the presence of additional mono-glycosylated (~77.4 kDa) and a-glycosylated (~75.2 kDa) isoforms (Fig. 2d).

PMM2-CDG and MPI-CDG (CDG–Ib, OMIM 602579) were excluded by enzymatic analysis of leukocytes and cultured skin fibroblasts and by direct sequence analysis of PMM2 and MPI genes.

Subsequent LLO analysis of fibroblasts of patient 1 showed no abnormal accumulation of any intermediate precursor.

DNA from patient 1 has further been analyzed to identify mutations by direct gene sequencing in the context of *ALG8-CDG*, *DPAGT1-CDG*, and *SRD5A3-CDG* (HGNC: 23161, 2995, and 25812, respectively). No defect in those genes could be identified in this patient.

Discussion

Since glycosylation of proteins occurs in all cell types, symptoms of a glycosylation deficiency are seen in multiple organs. Previous reports indicated that the diagnosis of CDG should be considered in each patient with hypotonia, dysmorphic features, and developmental delay accompanied by cataract (Morava et al. 2008). Our patients display multisystem symptoms; however, they also showed some characteristics that are not commonly seen in these pathologies. Interestingly, during a transient hepatic problem, patient 1 presented altered urine organic acids and plasma amino acids due to a mild liver dysfunction and a slightly increased serum galactose that became normal, with normal enzymes activities. The abnormalities in the lysosomal enzymes of these patients are compatible with the protein glycosylation defects that they have, which have been described in CDG patients.

In different reviews, CDG-I patients show ophthalmological problems. Common findings are bilateral cataract, glaucoma, and optic nerve atrophy. Infantile cataract might occur in various inborn errors of metabolism including galactosemia, peroxisomal disorders, and mitochondrial disorders. Cataract has been described in PMM2-CDG adult patients, but in children it is a rare clinical finding only described in a few ALG 8-CDG (CDG-Ih), ALG2-CDG (CDG-Ii), and in CDG-Ix patients (Thiel et al. 2003; Eklund et al. 2005; Morava et al. 2008); nevertheless, cataract is an important symptom in SRD5A3-CDG (Cantagrel et al. 2010 and Kahrizi et al. 2011).

In our patients, cataracts were present very early in life; the ophthalmological examination revealed that they had affected both eyes with opacities in the central portion of the lens as well as in the lens cortex (nuclear-cortical cataracts).

At present, patient 1 presents a severe phenotype and receives anti-epileptic medication, weekly physiotherapy, and assisted therapy, as well. She has made slight progress in her motor development, muscle tone, and social interaction, and continues to have involuntary movements of the head and the upper limbs. She has visual fixation and although she seems to recognize familiar faces, her communication is limited to undifferentiated vocalization.

As we can see from the literature that the severity of the disease has no correlation with a CDG I subtype, it seems that the disease course may be influenced by different factors that improve or worsen the patient's general condition (Morava et al. 2008). The clinical features of our patients were similar including axial hypotonia, psychomotor retardation, dysmorphic features, feeding problems, hepatomegaly, coagulopathy, and recurrent infections. Nevertheless, patient 2 was born with a severe liver involvement and recurrent neonatal infections that lead to an unfortunate early death; however, patient 1 developed a progressive neurological involvement after her first years of life.

Our patients present a similar phenotype to ALG1-CDG but not as severe as them, P1 is now 11 years old and her state is progressing. Compared with other CDG types, the phenotype in ALG1-CDG is very severe, with rapid development of microcephaly, seizures refractory to treatment, progressive stupor, and death in early infancy (Kranz et al. 2004). The hypoglycosylation of liver-derived serum glycoproteins is more profound than in other types of CDG-I. The IEF transferrin pattern in ALG1-CDG patients shows the major transferrin isoform almost missing, but P1 shows a profile that has either one or both carbohydrate side chains missing in about half of the transferrin population.

LLO analysis of fibroblasts did not identify abnormal accumulation of any dolichol-linked oligosaccharide precursors associated with some of the known forms of CDG, ruling out ALG6-CDG, ALG3-CDG, DPM1-CDG, MPDU1-CDG, ALG12-CDG, and ALG2-CDG. In this context, we screened *ALG8* and *DPAGT1* genes with a role in the assembly of the LLO in ER since a deficiency in those genes would also lead to normal LLO, in contrast to the other deficiencies described for the enzymes along this pathway. Finally, a normal LLO profile has recently been observed in patients with a defect in the polypropenol reductase, necessary for the synthesis of dolichol, due to mutations in the *SRD5A3* gene (Cantagrel et al. 2010). SRD5A3-CDG has also been excluded in our patients.

The results observed in our patients suggest studying the defects in dolichol synthesis up to the formation of Dol-PP-GlcNAc in addition to the recently described dolichol defects named DHDDS-CDG (dehydrodolichol diphosphate synthase deficiency) and DK1-CDG (dolichol kinase deficiency).

In conclusion, we could not identify the genetic deficiency in these patients yet. Further tests will be necessary to pinpoint the defect to identify the CDG type.

Acknowledgments The authors are grateful to Liesbeth Keldermans from the Center for Human Genetics and Sandra Van Aerschot from the Center for Metabolic Diseases for DNA analysis and capillary zone electrophoresis. The authors would like to thank Dr. Paz Briones, Dr. Rafaél Artuch, and Dr. María Antonia Vilaseca from Barcelona (Spain). At CEMECO, this work was supported by grants from CONICET (PIP6338), FONCyT – BID PICT2350, Catholic University of Córdoba, and Argentinean National Health Ministry.

Synopsis

This is a report of the clinical, biochemical, and molecular studies of two siblings characterized as CDG-Ix, identified in the context of a new program for the systematic identification of CDG in Argentina.

Author's Contributions

MBBM was responsible for preparative studies in terms of isoelectrofocusing, HPLC, CDG enzymatic assays, molecular studies, data gathering, and development of the manuscript.

AD and **NA** contributed in the biochemical studies.

NG is a physician specialized in metabolic disease who works at CEMECO.

LS and **DG** contributed in the mass spectrometry analysis.

GM was responsible for the molecular studies and contributed in the development and final draft of the manuscript.

JJ provided specialized opinion in the clinical diagnosis. He contributed in the development and final draft of the manuscript.

RDK provided specialized opinion and was responsible for the evaluation and the follow-up of the patient.

She coordinates the research in metabolic diseases in the Center for Study of Congenital Metabolopaties (CEMECO).

CGA is responsible for the research, conceived the idea for the chapter, and contributed to the development and final draft of the manuscript. She coordinated the CDG research program in CEMECO.

Competing Interest Statement

All authors confirm that they have no competing interests for declaration. The authors confirm independence from the sponsors. They did not receive any outside funding or grants in support of their research.

Details of Funding

This work was supported by grants from CONICET (PIP6338), (ANCyT-FONCyT) PICT2350, Catholic University of Córdoba, and Argentinean National Health Ministry. All authors confirm that they have no competing interests for declaration. The authors confirm independence from the sponsors; the content of the chapter has not been influenced by the sponsors.

Details of Ethics Approval

Written consent was obtained from the patients and their parents to participate in this study, and they allow us to submit this manuscript with images for the publication. A copy of written consent is available for review. The study was approved by the Ethics Committee of the Children's Hospital of Córdoba (CIEIS) Act Nº 95/2007. All studies were carried out in accordance with the Word Medical Association Declaration of Helsinki.

References

Artuch R, Ferrer I, Pinedo J et al (2003) Western blotting with diaminobenzidina detection for the diagnosis of congenital disorders of glycosilation. J Neurosci Methods 125:167–171

Babovic-Vuksanovic D, O'Brien J (2007) Laboratory diagnosis of congenital disorders of glycosylation type I by analysis of transferring glycoforms. Mol Diagn Ther 11:303–311

Barone R, Sturiale L, Garozzo D (2009) Mass spectrometry in the characterization of human genetic N-glycosylation defects. Mass Spectrom Rev 28:517–542

Cantagrel V, Lefeber D, Ng BG et al (2010) SRD5A3 is required for converting polyprenol to dolichol and is mutated in a congenital glycosylation disorder. Cell 142(2):203–17

de Lonlay P, Valayannopoulos V, Dupré T, Vuillaumier-Barrot S, Seta N (2008) Congenital disorders of glycosylation (CDG). Arch Pediatr 15:602–5

Denecke J, Kranz C, Von Kleist-Retzow J et al (2005) Congenital disorder of glycosylation type Id: clinical phenotype, molecular analysis, prenatal diagnosis, and glycosylation of fetal proteins. Pediatr Res 2:248–253

Eklund EA, Sun L, Westphal V, Northrop JL, Freeze HH, Scaglia F (2005) Congenital disorder of glycosylation (CDG)-Ih patient with a severe hepato-intestinal phenotype and evolving central nervous system pathology. J Pediatr 147:847–850

Guerrera I, Kleiner O (2005) Application of mass spectrometry in proteomics. Biosci Rep 25:71–93

Helander A, Husa A, Jeppsson J (2003) Improved HPLC method for carbohydrate-deficient transferrin in serum. Clin Chem 49:1881–1890

Jaeken J (2003) Congenital disorders of glycosilation (CDG): it's all in it! J Inherit Metab Dis 26:99–118

Jaeken J, Vanderschueren-Lodeweyckx M, Casaer P, SnoeckL CL, Eggermont E, Eeckels R (1980) Familial psychomotor retardation with markedly fluctuating serum prolactin, FSH and GH levels, partial TBG deficiency, increased serum arylsulphatase A and increased CSF protein: a new syndrome? Pediatr Res 14:179

Jaeken J, van Eijk HG, van der Heul C, Corbeel L, Eeckels R, Eggermont E (1984) Sialic acid-deficient serum and cerebrospinal fluid transferin in a newly reconized genetic syndrome. Clin Chim Acta 144:245–7

Jaeken J, Hennet T, Freeze HH, Matthijs G (2008) On the nomenclature of congenital disorders of glycosylation (CDG). J Inherit Metab Dis 31:669–672

Jaeken J, Hennet T, Matthijs G, Freeze HH (2009) CDG nomenclature: time for a change! Biochim Biophys Acta 1792:825–826

Kahrizi K, Hao Hu C, Garshasbi M et al (2011) Next generation sequencing in a family with autosomal recessive Kahrizi syndrome (OMIM 612713) reveals a homozygous frameshift mutation in SRD5A3. Eur J Hum Genet 19:115–117

Kranz C, Denecke J, Lehle L et al (2004) Congenital disorder of glycosylation type Ik (CDG-Ik): a defect of mannosyltransferase I. Am J Hum Genet 74:545–551

Lefeber D, Schonberg J, Morava E et al (2009) Deficiency of Dol-P-Man synthase subunit DPM3 bridges the congenital disorders of glycosylation with the dystroglycanopathies. Am J Hum Genet 85:76–86

Marklová E, Albahri Z (2007) Screening and diagnosis of congenital disorders of glycosylation. Clin Chim Acta 385:6–20

Matthijs G, Schollen E, Pardon E, Veiga-da-Cunha M, Jaeken J, Cassiman J-J, Van Schaftingen E (1997) Mutations in a phosphomannomutase gene, PMM2, on chromosome 16 in carbohydrate-deficient glycoprotein type I syndrome (Jaeken syndrome). Nat Genet 16:88–92, erratum: Nat Genet 16:316

Matthijs G, Schollen E, Van Schaftingen E, Cassiman JJ, Jaeken J (1998) Lack of homocygotes for the most frequent disease allele in carbohydrate-deficient glycoprotein syndrome type 1A. Am J Hum Genet 62:542–550

McKenzie F, Fietz M, Fletcher J, Smith R, Wright M, Jaeken J (2007) A previously undescribed form of congenital disorder of glycosylation with variable presentation in siblings: early fetal loss with hydrops fetalis, and infant death with hypoproteinemia. Am J Med Genet A 143A:2029–2034

Morava E, Wosik H, Karteszi J et al (2008) Congenital disorder of glycosylation type Ix: review of clinical spectrum and diagnostic steps. J Inherit Metab Dis 31:450–456

Quintana E, Navarro-Sastre A, Hernández-Pérez J et al (2009) Screening for congenital disorders of glycosylation (CDG): transferrin HPLC versus isoelectric focusing (IEF). Clin Biochem 42:408–15

Reinders J, Lewandrowski U, Moebius J, Wagner Y, Sickmann A (2004) Challenges in mass spectrometry-based proteomics. Proteomics 4:3686–3703

Schollen E, Dorland L, de Koning T et al (2000) Genomic organization of the human phosphomannose isomerase (MPI) gene and mutation analysis in patients with congenital disorders of glycosylation type Ib (CDG-Ib). Hum Mutat 16:247–252

Schollen E, Frank C, Keldermans L et al (2004) Clinical and molecular features of three patients with congenital disorders of glycosylation type Ih (CDG-Ih) (ALG8 deficiency). J Med Genet 41:550–556

Stibler H, Jaeken J (1990) Carbohydrate deficient serum transferrin in a new systemic hereditary syndrome. Arch Dis Child 65:107–111

Sturiale L, Barone R, Fuimara A et al (2005) Hypoglycosylation with increased fucosylation and branching of serum transferrin N-glycans in untrated galactosemia. Glycobiology 15:1268–76

Sturiale L, Barone R, Palmigiano A et al (2008) Multiplexed glycoproteomic analysis of glycosylation disorders by sequential yolk immunoglobulins immunoseparation and MALDI-TOF MS. Proteomics 8:3822–32

Thiel C, Schwarz M, Peng J, Grzmil M, Hasilik M, Braulke T (2003) A new type of congenital disorders of glycosylation (CDG Ii) provides new insights into the early steps of dolichol-linked oligosaccharide biosynthesis. J Biol Chem 278:22498–22505

van Schaftingen E, Jaeken J (1995) Phosphomannomutase deficiency is a cause of carbohydrate-deficient glycoprotein syndrome type I. FEBS Lett 377:318–320

Wu X, Rush JS, Karaoglu D et al (2003) Deficiency of UDP-GlcNAc: dolichol phosphate N-acetylglucosamine-1 phosphate transferase (DPAGT1) causes a novel congenital disorder of glycosylation type Ij. Hum Mutat 2:144–50

JIMD Reports
DOI: 10.1007/8904_2011_19

Received: 29 August 2010; Revised: 31 January 2011;
Accepted: 4 February 2011; Published online: 22 June
Communicated by: Gregory Pastores

Successful Screening for Gaucher Disease in a High-Prevalence Population in Tabuleiro do Norte (Northeastern Brazil): A Cross-Sectional Study

Rigoberto Gadelha Chaves, Janice Carneiro Coelho, Kristiane Michelin-Tirelli, Tibelle Freitas Maurício, Edineide de Freitas Maia Chaves, Paulo César de Almeida, Carlos Rômulo Filgueira Maurício, and Geraldo Barroso Cavalcanti Jr.

Abstract *Background:* Gaucher disease (GD) is a hereditary lysosomal storage disorder characterized by the accumulation of glucosylceramide, mainly in the cells of the reticuloendothelial system, due to a deficiency of the enzyme acid β-glucosidase (GBA). Diagnosis is usually based on measurement of GBA activity in peripheral leukocytes. The purpose of this study was to evaluate the ability of screening for GBA and chitotriosidase activity using dried blood spots on filter paper (DBS-FP) to identify individuals at high risk for GD in high-risk populations such as that of Tabuleiro do Norte, a small town in Northeastern Brazil.

Methods: Between 1 June 2007 and 31 May 2008, 740 consented residents and descendants of traditional families from Tabuleiro do Norte were submitted to screening with DBS-FP. Subjects with GBA activity < 2.19 nmol/h/mL were referred to the analysis of GBA and chitotriosidase activity in peripheral leukocytes and in plasma, respectively. Subjects at highest risk for GD (GBA activity in peripheral leukocytes < 5.6 nmol/h/mg protein) were referred to molecular analysis to confirm diagnosis.

Results: Screening with DBS-FP identified 135 subjects (18.2%) with GBA activity < 2.19 nmol/h/mL, 131 of whom remained in the study. In ten of these (7.6%), GBA activity in leukocytes was 2.6–5.5 nmol/h/mg protein. Subsequent molecular analysis confirmed six cases of heterozygosity and four normals for GD.

Conclusion: DBS-FP assay was shown to be an effective initial GD-screening strategy for high-prevalence populations in developing regions. Diagnosis could not be established from GBA activity in leukocytes alone, but required confirmation with molecular analysis.

Keywords Acid β-glucosidase · Diagnosis · Dried blood spots · Gaucher disease · Screening

Introduction

Gaucher disease (GD), the most common hereditary lysosomal storage disorder in humans (Beutler and Grabowski 2001), is caused by a deficiency of the enzyme acid β-glucosidase (GBA) (EC 3.2.1.45) (Brady et al. 1966) which leads to the accumulation of glucosylceramide, mainly in the cells of the reticuloendothelial system (Amaral et al. 2000; Charrow et al. 2000; Beutler and Grabowski 2001). More than 200 mutations have already been identified in the GBA gene (Abrahamov et al. 1995), justifying the heterogeneity of GD phenotypes. The three main clinical presentations are: type I – non-neuropathic form, accounting for 94% of cases; type II – acute neuropathic form (<1% of cases); and type III – subacute neuropathic form (5% of cases) (Charrow et al. 2000; Grabowski 2004).

R.G. Chaves (✉)
Programa de Pós-Graduação do Centro de Ciências da Saúde- UFRN, Rua Capitão José Rodrigues 4774, Centro, Tabuleiro do Norte, Ceará, Brazil
e-mail: rigobertogadelha@hotmail.com

J.C. Coelho
Department of Biochemistry, ICBS, UFRGS, Porto Alegre, Rio Grande do Sul, Brazil

K. Michelin-Tirelli
Department of Biochemistry, ICBS, UFRGS, Porto Alegre, Rio Grande do Sul, Brazil
and
Medical Genetics Center, HCPA, UFRGS, Porto Alegre, Rio Grande do Sul, Brazil

T.F. Maurício, E. de Freitas Maia Chaves, and C.R.F. Maurício
Municipal Department of Health, Rua Pe. Clicério, 4605 - Tabuleiro do Norte – Ceará – Brazil. CEP 62960-000

P.C. de Almeida
Health Sciences Center, UECE, Fortaleza, Ceará, Brazil

G.B. Cavalcanti Jr.
Departamento de Análises Clínicas e Toxicológicas - Faculdade de Farmácia, Programa de Pós-Graduação do Centro de Ciências da Saúde - UFRN, Avenida Gustavo Cordeiro de Farias S/N 1° Andar, Petrópolis, CEP: 59010-180, Natal-RN, Brazil
e-mail: gbcjunior@hotmail.com

The most common clinical manifestations of GD include anemia, thrombocytopenia, hepatosplenomegaly, and skeletal complications (pain crisis, bone injury, cortical and medullar bone infarction, medullar expansion, osteopenia, osteonecrosis, and pathological fractures) (Kaplan et al. 2006; Kishnani et al. 2009). Unfortunately, due to the low incidence and the variability of clinical manifestations of GD, many patients are misdiagnosed or remain undiagnosed (Mistry and Germain 2007).

GD is pan-ethnic. The incidence in the USA is estimated to be 1:40,000, with a higher prevalence among Ashkenazi Jews (1:400 to 1:800) (Grabowski 2004). In February 2009, 5,356 cases had been identified throughout the world, 531 of which in Brazil (ICGG Gaucher Registry 2009). Twenty of the Brazilian cases come from the State of Ceará (population: ~8.2 million), and, of these, as many as seven come from a small town (Tabuleiro do Norte) of only approximately 28,000 inhabitants. The high estimated prevalence of GD in Tabuleiro do Norte (1:4,000), probably the highest in Brazil, is believed to be due to a low migration rate and high level of inbreeding over many generations and possibly to the existence of Jewish heredity in some of the local families (Vieira 2000). Further analyses of the mutations of patients in Tabuleiro do Norte will help confirm this hypothesis.

The diagnosis of GD is mainly based on morphological findings (detection of Gaucher cells in tissues), enzyme activity, and molecular analysis (Beutler and Grabowski 2001). The definitive diagnosis requires the determination of GBA activity in peripheral leukocytes (Beutler and Saven 1990; Charrow et al. 1998) or in cultured fibroblasts from skin biopsies (Charrow et al. 1998; Beutler and Grabowski 2001).

High-risk populations may be screened with GBA and chitotriosidase activity assays using dried blood spots on filter paper (DBS-FP), followed by confirmatory tests in suspected cases (Chamoles et al. 2002; Civallero et al. 2006). DBS-FP is simple and time-saving, and samples are easy to transport. When combined with GBA activity assay in peripheral leukocytes, DBS-FP is a reliable and cost-effective strategy for public health-sponsored screening of lysosomal storage diseases in newborns and/or high-risk populations (Li et al. 2004; Meikle et al. 2004; Evans et al. 2005; Gelb et al. 2006). Moreover, the existence of effective treatment providing good control of the disease and considerable improvement in quality of life (Damiano et al. 1998; Masek et al. 1999; Giraldo et al. 2005; Weinreb et al. 2007) makes screening of high-risk populations imperative to diagnose new cases and initiate treatment when indicated (Civallero et al. 2006).

Some GD mutations are more strongly associated with disease severity and prognosis than others. Consequently, molecular diagnosis of GD is a very important aid in genetic counseling and population-screening programs for GD in high-risk groups. Molecular analysis makes it possible to identify homozygous and heterozygous subjects and subsidizes efforts to prevent the emergence of new cases in the population (Mistry et al. 1992; Abrahamov et al. 1995).

The purpose of this study was to evaluate the ability of screening for GBA activity with DBS-FP to identify individuals at high risk for GD in high-risk populations such as that of Tabuleiro do Norte, a small town in Northeastern Brazil.

Methods

Study Design and Population

The study was conducted in three stages: (1) evaluation of GBA and chitotriosidase activity in blood spots on filter paper; (2) evaluation of GBA and chitotriosidase activity in leukocytes and plasma, respectively; and (3) molecular analysis of samples from subjects highly suspected of GD.

This cross-sectional study included 740 residents and descendants of traditional families from Tabuleiro do Norte. The sample size was calculated with the formula for infinite populations. The level of statistical significance and sampling error was set at 5% and 3.6%, respectively. Due to lack of prior data, the total sample size was calculated from the 50% proportion.

The inclusion criteria for the study were: (1) being a resident and descendant of traditional families from Tabuleiro do Norte; (2) participation in health education sessions for GD; and (3) signing an informed consent form. Parents, siblings, and children of the seven patients who had a previously confirmed diagnosis of GD were excluded from the study. The study protocol was approved by the research ethics committee of *Hospital Geral César Cals* (Fortaleza, Brazil).

The population of Tabuleiro do Norte was informed of the study by the local media (newspapers and radio shows) and through educational lectures on GD. Thus, the recruitment of volunteers was nonrandom.

Laboratory Tests

Evaluation of GBA and Chitotriosidase Activity in Blood Spots on Filter Paper

All enzyme assays were performed at the Laboratory of Inborn Errors of Metabolism (Hospital das Clínicas, Porto Alegre, RS, Brazil). A 2-mL peripheral blood sample was collected from the forearm of each volunteer and placed in a heparin-coated test tube. Samples were homogenized, aspirated, separated into four parts, and dripped onto filter paper (903 Protein Saver Card, Whatman Inc., USA). Venipuncture was preferred over fingerstick to obtain a sample large enough for analysis. The filter paper was air-dried for 4 h, packed individually in a sealed plastic bag, and shipped at room temperature to the referral laboratory for measurement of GBA and chitotriosidase activity. Individual 3-mm disks containing approximately 3.6 μL of whole blood were incubated at 37°C with appropriate artificial substrates (4-methylumbelliferyl-β-D-glucoside and 4-methylumbelliferyl-*N*,*N*′,*N*″-triacetyl-β-chitotrioside) and dilution buffers. Samples were centrifuged and then submitted to fluorometric analysis (nanomoles of substrate hydrolyzed per hour per milliliter of blood) (Civallero et al. 2006).

Evaluation of GBA and Chitotriosidase Activity in Leukocytes and Plasma

New 10-mL blood samples were collected from subjects with GBA activity below 2.19 nmol/h/mL and stored in heparin-coated tubes. The samples were used to measure GBA activity in leukocytes (Peters et al. 1976) and chitotriosidase activity in plasma (Hollak et al. 1994). Leukocytes and plasma were separated immediately after collection, frozen, and shipped in dry ice to a referral laboratory 4,000 km away. The shipment was delivered within 48 h of collection.

The parameters for GBA activity in leukocytes used in the present study to classify the subjects in five groups (Table 1) were established by the referral laboratory based on enzyme analysis of samples from subjects previously diagnosed by molecular biology as homozygous, heterozygous, and normals for GD.

Molecular Analysis of Mutations

Samples from the subjects most suspected of GD (GBA in leukocytes < 5.6 nmol/h/mg protein) were screened for the major Brazilian mutations (N370S, L444P, G377S, and 55del) at the Department of Genetics and Evolutionary Biology of the Institute of Bio-Sciences (University of São Paulo, SP, Brazil). DNA was extracted from oral mucosa by the method of Richards et al. (1993). The mutations N370S, L444P, G377S, and 55del were analyzed with restriction fragment length polymorphism. After endonuclease digestion, fragments were submitted to 12% polyacrylamide gel electrophoresis (N370S and G377S) or 2% agarose gel electrophoresis (L444P and 55del) (Rozenberg et al. 2006).

Table 1 Classification adopted by the referral laboratory based on GBA activity in peripheral leukocytes

Classification	GBA activity in leukocytes (nmol/h/mg protein)
Probable GD	≤ 4.0
Borderline between probable GD and probable heterozygous	4.1–5.5
Probable heterozygous	5.6–9.9
Borderline between probable heterozygous and probable normal	10.0–16.4
Normal	≥ 16.5

When diagnosis was in doubt, samples from subjects with probable GD (GBA activity ≤4.0 nmol/h/mg protein) were shipped to the Molecular Development Laboratory of the Department of Pediatrics, University of Washington, USA, for sequence analysis of the entire coding region of the glucocerebrosidase gene (1q21) using dried blood on FTA filter paper. The use of FTA filter paper facilitates access of samples to diagnostic centers and thus provides an effective means of performing population-based mutational analysis of Gaucher disease internationally (Devost and Choy 2000). Sequencing will identify approximately 95% of mutations. It will often not identify deletions or rearrangements interfering with primer-binding sites.

Statistical Analysis

The collected data were submitted to descriptive analysis (frequency distribution and central tendency), followed by comparative analysis with the Mann–Whitney test and the chi-square test for categorical variables. Variables included gender, age, GBA activity in DBS-FP and in leukocytes, and chitotriosidase activity in DBS-FP and in plasma. The level of statistical significance was set at 5% ($p < 0.05$).

Subjects were initially distributed in two groups based on GBA activity in DBS-FP: (a) subjects at risk for GD and (b) normals (Fig. 1). The two groups were compared with regard to gender, age, and chitotriosidase activity. Samples from subjects at risk were subsequently submitted to analysis of GBA activity in leukocytes and classified in five categories according to criteria adopted by the referral laboratory (Table 1). Finally, these categories were compared with regard to chitotriosidase activity.

Results

From 1 June 2007 to 31 May 2008, 740 volunteers aged 31.4 ±19.2 years (range: 1–85), 496 (67.0%) of whom were female, were enrolled in the study. None of the participants had symptoms compatible with GD or were first-degree relatives of the seven residents of Tabuleiro do Norte known to have GD.

The mean GBA and chitotriosidase activity measured in the DBS-FP assay was 3.4±1.53 nmol/h/mL (range: 0.4–12.0) and 30.8±25.8 nmol/h/mL (range: 0.0–242.0), respectively. GBA levels were below the adopted cutoff value for risk of GD (2.19 nmol/h/mL) in 135 (18.2%) participants.

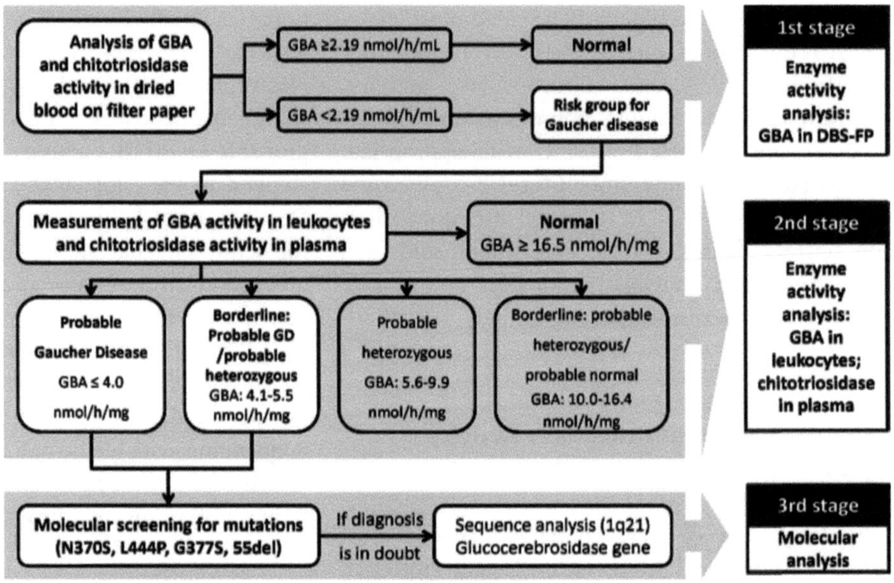

Fig. 1 Study design of screening for Gaucher disease in Tabuleiro do Norte (Ceará, Brazil)

Table 2 Comparison of gender, age, and chitotriosidase levels in DBS-FP between subjects at risk for GD and normals, according to the adopted cutoff value (<2.19 nmol/h/mL)

Variable	At risk for GD (GBA <2.19 nmol/h/mL), $n=135$	Normal (GBA ≥ 2.19 nmol/h/mL), $n=605$	p value
Female gender	87 (64.4%)	409 (67.6%)	>0.05[a]
Age (years)	32.0±19.2	31.4±19.3	>0.05[b]
Chitotriosidase (nmol/h/mL)	25.4±21.9	32.0±26.4	0.0014[b]

[a]Chi-square test
[b]Mann–Whitney test

Table 3 Frequency distribution of GBA activity in peripheral leukocytes of 131 subjects with suspected GD, according to the classification adopted by the referral laboratory

Classification	N (%)	CI 95%	Chitotriosidase (nmol/h/mL ± SD)
Probable GD	4 (3.1%)	0.8–7.6%	82.75±13.5
Borderline between probable GD and probable heterozygous	6 (4.6%)	1.7–9.7%	61.47±72.53
Probable heterozygous	63 (48.1%)	39.3–57.0%	43.37±36.0
Borderline between probable heterozygous and probable normal	55 (42.0%)	33.4–50.9%	57.89±69.54
Normal	3 (2.3%)	0.5–6.5%	42.33±41.54

Subjects below and above cutoff did not differ significantly with respect to gender or age, but chitotriosidase activity in DBS-FP was significantly lower among the former (Table 2).

Four of the 135 subjects at risk for GD did not wish to participate in the second part of the study. New blood samples were collected from the remaining 131 subjects to determine GBA activity in peripheral leukocytes and chitotriosidase activity in plasma. Four subjects (3.1%) with GBA levels in the range 2.6–4.0 nmol/h/mg protein were classified as "probable GD," 6 (4.6%) were "borderline between probable GD and probable heterozygous," 63 (48.1%) were classified as "probable heterozygous," and 55 (42.0%) were "borderline between probable heterozygous and probable normal." Only three (2.3%) displayed normal GBA activity levels (Table 3). Subjects with probable GD did not differ significantly from subjects in the four other GBA ranges with regard to gender and age ($p > 0.05$) but displayed higher plasma chitotriosidase levels (82.75± 13.5 nmol/h/mL vs. 50.49±54.80 nmol/h/mL, $p = 0.023$).

The GBA and chitotriosidase activity assays were repeated for the ten subjects in the two lowest ranges of GBA activity, confirming the initial results. The same subjects were submitted to additional evaluations, including clinical examinations, laboratory tests (complete peripheral blood count and aminotransferase dosage), X-ray scanning of pelvis, femurs, lumbar spine and chest, abdominal ultrasonography, magnetic resonance imaging of femurs, and bone densitometry of the lumbar spine. To further confirm the diagnosis, samples of oral mucosa were submitted to molecular screening for the four major Brazilian GD mutations (N370S, L444P, G377S, and 55del). Four subjects presented none of the mutations, but six subjects were found to be heterozygous (G377S/-). Complete individual genetic counseling was given upon study completion.

The four subjects with GBA activity in leukocytes below 4.1 nmol/h/mg protein (two heterozygous male siblings, mutation G377S/-, 12 and 17 years of age with GBA activity of 4.0 and 2.6 nmol/h/mg protein, respectively, and two unrelated females aged 39 and 44 years with GBA activity of 4.0 nmol/h/mg protein) were asymptomatic and presented no significant clinical changes in laboratory or imaging tests. Thus, to test for mutations other than N370S, L444P, G377S, and 55del, dried blood samples on FTA filter paper were collected and submitted to sequence analysis of the entire coding region of the glucocerebrosidase gene (1q21). The DNA analysis confirmed the diagnosis of two heterozygous subjects (p.G377S/wt) and two normals (wt/wt).

Discussion

The results of our study show that GBA activity assay in DBS-FP is an efficient screening strategy for the identification of subjects at risk of GD in high-prevalence populations. It also draws attention to the risk of false-positive diagnosis of GD in asymptomatic subjects when the measurement of GBA activity in leukocytes is used as the sole criterion. Molecular analysis was indispensable in this study to clarify doubts and confirm diagnosis. Only one mutation (G377S) was detected among the ten subjects at highest risk for GD. The fact that G377S is the third-most common mutation in subjects with GD in Portugal and Spain (Amaral et al. 1996) and rare in Ashkenazi Jews suggests that the mutation observed in Tabuleiro do Norte may be due to Portuguese ancestry. Further studies are necessary to clarify this issue.

The initial GBA activity assay in DBS-FP identified 135 of 740 subjects to be at risk for GD. The adopted cutoff value (<2.19 nm/h/mL) was above the value recommended by Civallero et al. (2006) (1.78 nmol/h/mL) to reduce the number of false-negative results and ensure the identification of all potential GD patients and carriers.

Four of the 135 patients selected in the initial screening withdrew from the study. Of the remaining 131 subjects, 4 were classified as "probable GD" (0.54%; CI 95%; 0.2–1.5%) and 124 as "possible heterozygous" (16.8%; CI 95%; 14.2–19.7%). The level of GBA activity was normal for the remaining 608 subjects (605 in DBS-FP, 3 in leukocytes) (82.6%; CI 95%; 79.6–85.2%). Thus, to identify one case of probable GD, 185 apparently healthy subjects had to be screened, and to identify one case of probable heterozygosity, only six subjects needed to be examined. In a high-risk population such as that of Tabuleiro do Norte, correct identification of affected individuals and carriers, including molecular screening for mutations, is essential to clarify doubts about diagnosis and provide adequate genetic counseling and, consequently, to reduce the incidence and prevalence of GD.

The fact that chitotriosidase activity in plasma was significantly higher in subjects with GBA activity in leukocytes below 4.1 nmol/h/mg protein than in subjects classified as "probable heterozygous" or "normal" ($p < 0.05$) matches the results published by other researchers (Hollak et al. 1994; Aerts et al. 2003; Cabrera-Salazar et al. 2004; Schoonhoven et al. 2007). Surprisingly, chitotriosidase activity in DBS-FP was lower for subjects with GBA activity <2.19 nmol/h/mL in DBS-FP than for normal subjects ($p < 0.05$), suggesting that, at least in this study, chitotriosidase activity was not a useful marker of GD. The discrepancy between the measures of chitotriosidase activity in DBS-FP and in plasma is difficult to explain, but the results of one or both of the chitotriosidase assays may have been affected by unfavorable sample shipping conditions.

The recruitment process was the most important limitation of this study. Since the selection of participants was nonrandom and because relatives of previously diagnosed patients were likely to be particularly interested in the study, the prevalence of homozygotes and heterozygotes in the study population may not have been representative of the general population. Unfortunately, it was not possible to completely avoid recruitment bias in the study.

Screening with enzyme assays instead of molecular analysis may at first sight appear a limitation. However, molecular analysis was not an option for initial screening due to lack of previous knowledge of the genetic profile of the population and due to questions of cost-effectiveness and infrastructure (the study was conducted in a socioeconomically challenged region). Despite its apparent limitations, we believe that triage with DBS-FP may even be used to improve the cost-effectiveness of screening with molecular analysis.

To our knowledge, this study represents the largest population screening for GD in Brazil using DBS-FP. Initial screening with DBS-FP, followed by a confirmatory GBA activity assay in peripheral leukocytes and molecular analysis, was shown to be an effective GD-screening strategy for high-prevalence populations in developing regions. Further studies are needed to determine the cost-effectiveness of different methods, alone and in combination, used to screen for GD in other high-prevalence populations.

Acknowledgments The authors would like to thank the team at Genzyme do Brasil, sponsor of the study. Also many thanks to Dr. Elisa Sobreira for scientific support, Prof. Antonio Toledo Jr. for assistance with the manuscript, and Dr. Erlane Marques Ribeiro for help with the study design. We are also in debt to João Márcio da Silva, head of the Municipal Department of Health, and to the staff of the Family Health Program in Tabuleiro do Norte for their cooperation. Last but not least, we are appreciative of the valuable support received from ACDG (Associação Cearense de Profissionais atuantes em Doenças Genéticas, Pacientes, Familiares e Voluntários).

Synopsis

Enzyme assay with DBS-FP was shown to be an effective initial GD-screening strategy for high-prevalence populations in developing regions when followed by measurement of GBA activity in leukocytes, but molecular analysis was necessary to confirm diagnosis (homozygous, heterozygous, and normals for GD).

Conflicts of Interest

Rigoberto Chaves and Tibelle Maurício received educational grants from Genzyme do Brasil to help develop the study. Genzyme do Brasil covered the travel expenses and training of Rômulo Maurício and sponsored the laboratory tests performed by Janice Coelho and Kristiane Michelin-Tirelli. No other conflicts of interest are reported.

References

Abrahamov A, Elstein D et al (1995) Gaucher's disease variant characterised by progressive calcification of heart valves and unique genotype. Lancet 346(8981):1000–1003

Aerts JM, Hollak C et al (2003) Biochemistry of glycosphingolipid storage disorders: implications for therapeutic intervention. Philos Trans R Soc Lond B Biol Sci 358(1433):905–914

Amaral O, Pinto E, Fortuna M, Lacerda L (1996) Sa Miranda MC. Type 1 Gaucher disease: identification of N396T and prevalence of glucocerebrosidase mutations in the Portuguese. Hum Mutat 8:280–281

Amaral O, Marcao A et al (2000) Gaucher disease: expression and characterization of mild and severe acid beta-glucosidase mutations in Portuguese type 1 patients. Eur J Hum Genet 8(2):95–102

Beutler E, Grabowski G (2001) Gaucher disease. In: Scriver C, Beaudet A, Sly W, Valle D (eds) The metabolic and molecular bases of inheredited disease. McGraw-Hill, New York, pp 3635–3668

Beutler E, Saven A (1990) Misuse of marrow examination in the diagnosis of Gaucher disease. Blood 76(3):646–648

Brady RO, Kanfer JN et al (1966) Demonstration of a deficiency of glucocerebroside-cleaving enzyme in Gaucher's disease. J Clin Invest 45(7):1112–1115

Cabrera-Salazar MA, O'Rourke E et al (2004) Correlation of surrogate markers of Gaucher disease. Implications for long-term follow up of enzyme replacement therapy. Clin Chim Acta 344(1–2):101–107

Chamoles NA, Blanco M et al (2002) Gaucher and Niemann-Pick diseases–enzymatic diagnosis in dried blood spots on filter paper: retrospective diagnoses in newborn-screening cards. Clin Chim Acta 317(1–2):191–197

Charrow J, Esplin JA et al (1998) Gaucher disease: recommendations on diagnosis, evaluation, and monitoring. Arch Intern Med 158(16): 1754–1760

Charrow J, Andersson HC et al (2000) The Gaucher registry: demographics and disease characteristics of 1698 patients with Gaucher disease. Arch Intern Med 160(18):2835–2843

Civallero G, Michelin K et al (2006) Twelve different enzyme assays on dried-blood filter paper samples for detection of patients with selected inherited lysosomal storage diseases. Clin Chim Acta 372(1–2):98–102

Damiano AM, Pastores GM et al (1998) The health-related quality of life of adults with Gaucher's disease receiving enzyme replacement therapy: results from a retrospective study. Qual Life Res 7(5):373–386

Devost NC, Choy FY (2000) Mutation analysis of Gaucher disease using dot-blood samples on FTA filter paper. Am J Med Genet 94(5):417–420

Evans MI, Galen RS et al (2005) Principles of screening. Semin Perinatol 29(6):364–366

Gelb MH, Turecek F et al (2006) Direct multiplex assay of enzymes in dried blood spots by tandem mass spectrometry for the newborn screening of lysosomal storage disorders. J Inherit Metab Dis 29(2–3):397–404

Giraldo P, Solano V et al (2005) Quality of life related to type 1 Gaucher disease: Spanish experience. Qual Life Res 14(2):453–462

Grabowski GA (2004) Gaucher disease: lessons from a decade of therapy. J Pediatr 144(5 Suppl):S15–S19

Hollak CE, van Weely S et al (1994) Marked elevation of plasma chitotriosidase activity. A novel hallmark of Gaucher disease. J Clin Invest 93(3):1288–1292

ICGG Gaucher Registry (2009) Relatório do Brasil: Brasil comparado ao resto do mundo. Relatório Anual de 2009

Kaplan P, Andersson HC et al (2006) The clinical and demographic characteristics of nonneuronopathic Gaucher disease in 887 children at diagnosis. Arch Pediatr Adolesc Med 160(6):603–608

Kishnani PS, DiRocco M et al (2009) A randomized trial comparing the efficacy and safety of imiglucerase (Cerezyme) infusions every 4 weeks versus every 2 weeks in the maintenance therapy of adult patients with Gaucher disease type 1. Mol Genet Metab 96(4):164–170

Li Y, Scott CR et al (2004) Direct multiplex assay of lysosomal enzymes in dried blood spots for newborn screening. Clin Chem 50(10):1785–1796

Masek BJ, Sims KB et al (1999) Quality of life assessment in adults with type 1 Gaucher disease. Qual Life Res 8(3):263–268

Meikle PJ, Ranieri E et al (2004) Newborn screening for lysosomal storage disorders: clinical evaluation of a two-tier strategy. Pediatrics 114(4):909–916

Mistry P, Germain DP (2007) Therapeutic objectives in Gaucher disease. Rev Méd Interne 28(Suppl 2):S171–S175

Mistry PK, Smith SJ et al (1992) Genetic diagnosis of Gaucher's disease. Lancet 339(8798):889–892

Peters S, Coyle P et al (1976) Differantiation of beta glucocerebrosidase form beta glucosidase in human tissues using sodium taurocholate. Arch Biochem Biophys 175:562–569

Richards B, Skoletsky J, Shuber AP, Balfour R, Stern RC, Dorkin HL, Parad RB, Witt D, Klinger KW (1993) Multiplex PCR amplification from the CFTR gene using DNA prepared from buccal brushes/swabs. Hum Mol Genet 2:159–163

Rozenberg R, Araujo FT et al (2006) High frequency of mutation G377S in Brazilian type 3 Gaucher disease patients. Braz J Med Biol Res 39(9):1171–1179

Schoonhoven A, Rudensky B et al (2007) Monitoring of Gaucher patients with a novel chitotriosidase assay. Clin Chim Acta 381(2): 136–139

Vieira V (2000) Os Gadelhas no Mundo. http://www.gentree.org.br/artigos/gadelha.htm. Accessed 13 May 2009

Weinreb N, Barranger J et al (2007) Imiglucerase (Cerezyme) improves quality of life in patients with skeletal manifestations of Gaucher disease. Clin Genet 71(6):576–588

JIMD Reports
DOI: 10.1007/8904_2011_20

Received: 14 September 2010; Revised: 16 January 2011;
Accepted: 20 January 2011; Published online: 22 June
Communicated by: Alberto Burlina

Two Cases of Pulmonary Hypertension Associated with Type III Glycogen Storage Disease

Teresa M. Lee, Erika S. Berman-Rosenzweig, Alfred E. Slonim, and Wendy K. Chung

Abstract Glycogen storage diseases (GSDs) comprise a large, heterogeneous group of disorders characterized by abnormal glycogen deposition. Multiple cases in the literature have demonstrated an association between GSD type I and pulmonary arterial hypertension (PAH). We now also report on two patients with GSD type III and PAH, a novel association. The first patient was a 16-year-old girl of Nicaraguan descent with a history of hepatomegaly and growth retardation. Molecular testing identified a homozygous 17delAG mutation in *AGL* consistent with GSD type IIIb. At the age of 16, she was found to have PAH and was started on medical therapy. Two years later, she developed acute chest pain and died shortly thereafter. The second patient is a 13-year-old girl of Colombian descent homozygous for the c.3911dupA mutation consistent with GSD IIIa. An echocardiogram at age 2 showed left ventricular hypertrophy, which resolved following the institution of a high protein, moderate carbohydrate diet during the day and continuous gastric-tube feeding overnight. At the age of 12, she was found to have pulmonary hypertension. She was started on sildenafil, and her clinical status has shown marked improvement including normalization of her elevated transaminases. PAH may be a rare association in patients with GSD IIIa and IIIb and should be evaluated with screening echocardiograms for cardiac hypertrophy or if they present with symptoms of right-sided heart failure such as shortness of breath, chest pain, cyanosis, fatigue, dizziness, syncope, or edema. Early diagnosis of PAH is important as increasingly effective treatments are now available.

Keywords Amylo-1,6-glucosidase · Genetic · Hepatomegaly · Metabolic

Abbreviations

AGL	Amylo-1,6-glucosidase
GSD	Glycogen storage disease
PAH	Pulmonary arterial hypertension
VSD	Ventricular septal defect

Introduction

Glycogen storage diseases (GSDs) comprise a large, heterogeneous group of disorders characterized by abnormal glycogen deposition. GSD type III (OMIM 232400), characterized by deficiency of glycogen debrancher enzyme or amylo-1,6-glucosidase (AGL; EC 3.2.1.33), is further subdivided into groups including IIIa with hepatic and muscle involvement and IIIb an isolated hepatic form (Shin 2006).

Pulmonary arterial hypertension (PAH) is a rare disorder, with an estimated incidence of two to three per million per year (Gaine and Rubin 1998; Humbert et al. 2006), characterized by sustained increase in mean pulmonary artery pressure (above 25 mmHg at rest), a normal pulmonary capillary wedge pressure, and increased pulmonary vascular resistance. Before the advent of modern therapies, life expectancy for adults with idiopathic PAH was less than 3 years from diagnosis and less than 10 months for children (D'Alonzo et al. 1991). PAH may be heritable (HPAH), idiopathic (IPAH), associated with either drug or toxin exposures (fenfluramine derivatives) or other medical conditions including connective tissue diseases, HIV infection, congenital heart disease, sickle cell disease, and portal hypertension.

T.M. Lee, E.S. Berman-Rosenzweig, and A.E. Slonim
Department of Pediatrics, Columbia University Medical Center, New York, NY, USA

W.K. Chung (✉)
Department of Pediatrics, Columbia University Medical Center, New York, NY, USA
and
Russ Berrie Medical Science Pavilion, 1150 St. Nicholas Avenue, Room 620, New York, NY 10032, USA
e-mail: wkc15@columbia.edu

Multiple cases have been described in the literature that demonstrate an association between GSD type I and severe pulmonary hypertension (Humbert et al. 2002b). To date, there have been no reported cases of GSD type III with PAH, although GSD III shares many clinical features with GSD I.

We present two patients, one with GSD IIIa and the other with GSD IIIb, and pulmonary hypertension.

Case Reports

The first patient was a 16-year-old girl of Nicaraguan descent who first presented to our pulmonary hypertension center with a 2-year history of progressive dyspnea on exertion with occasional cyanosis. At 8 months of age, she was noted to have hepatomegaly, but a liver biopsy performed at that time was reportedly normal. From the age of 11 months, there was evidence of growth retardation with both height and weight below the third percentile, but no further evaluation was undertaken.

On presentation at 16 years, a 2D echocardiogram was performed which demonstrated a dilated right ventricle with moderate right ventricular hypertrophy and slightly reduced right ventricular function. Her estimated right ventricular systolic pressure was 78 mmHg. The pulmonary regurgitation gradient was 36 mmHg. There was posterior bowing of the interatrial and interventricular septum consistent with at least systemic pulmonary arterial pressure. There was right to left shunting through a small interatrial communication.

Cardiac catheterization was performed and demonstrated elevated mean pulmonary artery pressure (mean of 44 mmHg) with a normal mean pulmonary capillary wedge pressure (Table 1). Her cardiac index was low with extremely elevated pulmonary vascular resistance index at 24 u*m^2. There was no significant acute response to the administration of 100% oxygen via face mask, inhaled nitric oxide at 80 ppm, or intravenous epoprostenol at a maximum dose of 10 ng/kg/min. She was evaluated for causes of PAH and initially had elevated TSH and T4 that normalized on several serial repeat measurements.

Table 1 Cardiac catheterization data from patients 1 and 2 with normal reference values show elevated mean pulmonary artery pressure, normal mean capillary wedge pressure, and increased pulmonary vascular resistance consistent with the diagnosis of pulmonary hypertension in both patients. All data represent initial testing done on room air

	Patient 1	Patient 2	Normal values
Systemic blood pressure (mmHg)	110/68	83/50	109–114/64–68
Mean right atrial pressure (mmHg)	14	8	2–6
Pulmonary artery pressure (mmHg)	88/22	69/37	15–25/8–12
Mean pulmonary artery pressure (mmHg)	44	51	<20
Mean capillary wedge pressure (mmHg)	6	14	<15
Cardiac index (L/min/m^2)	1.6	2.9	3–5
Pulmonary vascular resistance index (u*m^2)	24	9	3

Liver biopsy was repeated at the age of 16 when she presented with pulmonary hypertension and marked hepatomegaly, which appeared out of proportion to her pulmonary hypertension. Liver biopsy showed marked glycogen storage affecting the hepatocytes with perisinusoidal and portal fibrosis and developing nodularity consistent with GSD. Further analysis of the liver showed elevated glycogen content with short outer branches and absence of debrancher activity consistent with type III GSD. Molecular testing showed that she had a homozygous 17delAG mutation in *AGL* consistent with GSD type IIIb. She was treated with albuterol, digoxin, furosemide, spironolactone, and continuous intravenous epoprostenol with some stabilization in her clinical symptoms.

At the age of 18, she presented to our emergency department with acute chest pain. She was tachycardic (heart rate 164 bpm) and tachypneic (respiratory rate 40 breaths/min), with a systemic arterial oxygen saturation level of 80%. One month before this presentation, her thyroid function tests were markedly abnormal with positive antithyroid peroxidase antibodies. Subsequent radioactive iodine scanning confirmed the diagnosis of Graves' disease. She was treated with propylthiouracil. She was admitted to the pediatric intensive care unit where she was found to be severely hypoglycemic with a glucose level of 6 mg/dL (normal 50–110 mg/dL) and to be hyperkalemic (8.6 mM/L; normal 3.6–5 mM/L). She was stabilized and put on BiPAP of 10/5 with inhaled nitric oxide at 20 ppm. Less than 10 h later, she complained of severe chest pain and became asystolic. Despite resuscitation attempts, she died.

The second patient is a 13-year-old girl of Colombian descent who was first noted to have poor weight gain and a large abdomen as an infant. A liver biopsy was performed at 14 months, which demonstrated glycogen accumulation, and she was diagnosed with GSD type III. Genetic testing of the *AGL* gene showed that she is homozygous for the c.3911dupA mutation consistent with GSD IIIa. An echocardiogram at age 2 showed left ventricular dilation (left ventricular end-diastolic dimension 3.28 cm, $z = 2.90$) with some hypertrophy in the presence of a large subaortic ventricular septal defect (VSD) with a large systemic to pulmonary shunt. She underwent repair of the VSD at 2 years of age, with a small residual restrictive VSD. Cardiac biopsy performed at the time of VSD repair showed subendocardial vacuolization of myocytes due to abnormal cytoplasmic glycogen deposits. Following the institution of therapy consisting of a high protein, moderate carbohydrate diet during the day, and continuous gastric-tube feeding overnight

(Slonim et al. 1982), the cardiac hypertrophy resolved and her clinical status progressively improved. Continuous normoglycemia was maintained and normal developmental milestones were achieved, with normal height, weight, and pubertal development. At the age of 12, she complained of shortness of breath and had an echocardiogram demonstrating tricuspid regurgitation jet. She then underwent cardiac catheterization and was found to have pulmonary hypertension. Resting pulmonary artery pressure was more than two-thirds of the systemic arterial pressure with an elevated mean right atrial pressure. Mean pulmonary capillary wedge pressure was normal. There was a mild response to the administration of 100% oxygen with the pulmonary artery pressure decreasing by 10 mmHg. In addition, when nitric oxide was administered at 80 ppm, her mean pulmonary arterial pressure fell from 52 to 35 mmHg. At rest, her pulmonary vascular resistance index was high at 9 u*m^2 which decreased to 5 u*m^2 with acute vasodilator testing. Superior vena cava saturation was 83% with no step up to the pulmonary artery which was 75%, with a Qp:Qs of 1.6 indicating no significant left to right shunt through the small residual VSD. Therefore, her pulmonary hypertension could not be attributed to the residual VSD. She was started on sildenafil 20 mg p.o. three times daily for PAH, and her clinical status has shown marked improvement. In addition, she had a history of elevated transaminase levels in the range of AST 1,042 U/L and ALT 813 U/L (normal 12–38, and 7–41 U/L, respectively) with no other evidence of synthetic liver dysfunction. Following treatment with sildenafil, the transaminase levels decreased to 149 U/L (AST) and 147 U/L (ALT) within 3 months and have normalized, 37 U/L (AST) and 34 U/L (ALT), after 3 years of sildenafil therapy. Interestingly, there was no overt clinical evidence of right heart failure or right ventricular systolic dysfunction by noninvasive testing before medical treatment with sildenafil to explain the decline in transaminases following treatment.

Discussion

While GSD type I has been repeatedly associated with pulmonary hypertension, there are no reports in the literature of PAH associated with type III GSD. The underlying connection between PAH in GSD I and GSD III remains to be elucidated and should be independently confirmed. The fact that we observed PAH with both GSD type IIIa and IIIb suggests that cardiac involvement is not required for PAH since cardiomyopathy may occur in type IIIa, but is not observed in type IIIb GSD.

In one study, patients with type Ia GSD were found to have elevated serotonin levels as did individuals with PAH when compared to age-matched controls (Humbert et al. 2002a). Moreover, one individual with both type Ia GSD and PAH had dramatically elevated plasma serotonin concentrations. Since not all patients with GSD I or GSD III go on to develop PAH, there must be additional factors required.

We hypothesize that there could be a diffusible substance produced by or not cleared by the diseased liver that causes pulmonary vasoconstriction. This is one mechanism hypothesized in cases of portopulmonary hypertension in which portosystemic collaterals allow humoral substances, normally metabolized by the liver, access to the pulmonary circulation (Panos and Baker 1996). Alternatively, there could be an intrinsic problem in the underlying pulmonary vasculature due to glycogen accumulation. As in portopulmonary hypertension, lung pathology from patients with type I GSD seems to consistently demonstrate pathological changes that are indistinguishable from those seen in classic idiopathic PAH (Humbert et al. 2002b; Pizzo 1980).

In conclusion, while PAH may be a rare association, the diagnosis should be considered in patients with GSD III if they present with cardiorespiratory symptoms such as shortness of breath, chest pain, cyanosis, fatigue, dizziness, syncope, or edema. Moreover, cardiac involvement in GSD is well established, and screening echocardiograms are recommended to examine at wall thickness, ventricular mass, in addition to systolic and diastolic function in GSD III patients (Kishnani et al. 2010). At the time of screening echocardiogram, we suggest examination for signs of elevated pulmonary arterial pressure such as right atrial or right ventricular enlargement, systolic flattening or posterior bowing of the interventricular septum, and measurement of the tricuspid regurgitation jet. Therapeutic drugs including prostacyclin, endothelin-receptor antagonists, and phosphodiesterase inhibitors, like sildenafil, have shown promising results in patients, including children and infants, with PAH (Suesaowalak et al. 2010). Recognizing elevated pulmonary arterial pressure early can be especially important, as increasingly effective treatments are now available for PAH.

Synopsis

Pulmonary hypertension association with glycogen storage disease IIIa and IIIb.

References to Electronic Database

Glycogen storage disease type III: OMIM 232400. Amylo-1,6-glucosidase: EC 3.2.1.33. HUGO-approved gene symbol: *AGL*.

References

D'Alonzo GE, Barst RJ, Ayres SM et al (1991) Survival in patients with primary pulmonary hypertension. Results from a national prospective registry. Ann Intern Med 115:343–349

Gaine SP, Rubin LJ (1998) Primary pulmonary hypertension. Lancet 352:719–725

Humbert M, Labrune P, Sitbon O et al (2002a) Pulmonary arterial hypertension and type-I glycogen-storage disease: the serotonin hypothesis. Eur Respir J 20:59–65

Humbert M, Labrune P, Simonneau G (2002b) Severe pulmonary arterial hypertension in type 1 glycogen storage disease. Eur J Pediatr 161(Suppl 1):S93–S96

Humbert M, Sitbon O, Chaouat A et al (2006) Pulmonary arterial hypertension in France: results from a national registry. Am J Respir Crit Care Med 173:1023–1030

Kishnani PS, Austin SL, Arn P et al (2010) Glycogen storage disease type III diagnosis and management guidelines. Genet Med 12(7): 446–463

Panos RJ, Baker SK (1996) Mediators, cytokines, and growth factors in liver-lung interactions. Clin Chest Med 17:151–169

Pizzo CJ (1980) Type I glycogen storage disease with focal nodular hyperplasia of the liver and vasoconstrictive pulmonary hypertension. Pediatrics 65:341–343

Shin YS (2006) Glycogen storage disease: clinical, biochemical, and molecular heterogeneity. Semin Pediatr Neurol 13:115–120

Slonim AE, Weisberg C, Benke P et al (1982) Reversal of debrancher deficiency myopathy by the use of high-protein nutrition. Ann Neurol 11:420–422

Suesaowalak M, Cleary JP, Chang AC (2010) Advances in diagnosis and treatment of pulmonary arterial hypertension in neonates and children with congenital heart disease. World J Pediatr 6:13–31

JIMD Reports
DOI: 10.1007/8904_2011_21

Received: 6 October 2010; Revised: 24 January 2011;
Accepted: 25 January 2011; Published online: 22 June
Communicated by: Johan Van Hove

Favorable Long-Term Outcome Following Severe Neonatal Hyperammonemic Coma in a Patient with Argininosuccinate Synthetase Deficiency

Isabelle De Bie, Emmanuelle Lemyre, and Marie Lambert

Abstract This chapter reports on the sequelae-free 8-year follow-up with normal growth, intellectual development, and schooling of a boy with argininosuccinate synthetase deficiency (citrullinemia type I) who was rescued from severe neonatal hyperammonemic coma at 8 days of life (peak ammonia level of 1,058 μmol/L). Important clinical management aspects were: rapidity of response to emergency therapeutic measures that included specific drug regimen, protein restriction, optimal caloric intake and hemodialysis, short coma duration (14 h), possible neuroprotective effect of mild systemic hypothermia during the acute episode, long-term metabolic control with strict compliance to standard of care therapeutic and dietary regimens, active prevention of subsequent hyperammonemic episodes, and early neurodevelopmental evaluations and interventions. We conclude that good long-term neurological outcome following rescue from neonatal hyperammonemic coma is rarely reported but attainable. Prospective registries and interventional studies regrouping clinical data from urea cycle disorders patients will assist clinicians in instituting the appropriate therapeutic measures to provide the best prospect of positive long-term outcome for these children.

I. De Bie
Medical Genetics Division, Department of Pediatrics, Centre Hospitalier Universitaire Sainte-Justine, Université de Montréal, 3175 Côte Sainte-Catherine, Montréal, QC, Canada H3T-1C5
and
Medical Genetics Department, Hôpital de Montréal Pour Enfants, Centre Universitaire de Santé McGill, 2300 rue Tupper, Montréal, QC, Canada H3H 1P3
e-mail: Isabelle.debie@muhc.mcgill.ca

E. Lemyre, and M. Lambert (✉)
Medical Genetics Division, Department of Pediatrics, Centre Hospitalier Universitaire Sainte-Justine, Université de Montréal, 3175 Côte Sainte-Catherine, Montréal, QC, Canada H3T-1C5
e-mail: Emmanuelle.lemyre@recherche-ste-justine.qc.ca, Marie.lambert@umontreal.ca

Keywords Argininosuccinate synthetase · Hyperammonemia · Inborn errors of metabolism · Neurodevelopment · Single gene disorder

Abbreviations

ASS Argininosuccinate synthetase
UCD Urea cycle defects

Introduction

Urea cycle defects (UCD) are inherited metabolic disorders leading to abnormal ammonia accumulation. The typical clinical picture associated with severe defects is that of a newborn infant, who, after an uneventful pregnancy and immediate neonatal course, becomes rapidly poorly responsive and evolves toward catastrophic neurological deterioration with seizures, coma, and death. UCD may also present later in life, either as a severe acute episode or as a more chronic course of developmental delay, failure to thrive, and a combination of neurological and digestive manifestations. Treatment options include protein restriction to limit exogenous sources of nitrogen, ammonia clearance through nitrogen-scavenging medications, supplementation with urea cycle intermediates (L-arginine or L-citrulline), and dialysis during acute events. Liver transplantation presents a curative option to achieve long-term metabolic remission (Leonard and McKiernan 2004; Meyburg and Hoffmann 2010).

Access to these treatments has significantly improved the survival of infants with UCD, but the neurological prognosis for survivors of acute hyperammonemic episodes remains uncertain. Children who survive the initial hyperammonemic crisis are at significant risk of disabling neurological sequelae (Msall et al. 1984; Saudubray et al. 1999; Bachmann 2003a; Gropman and Batshaw 2004; Nassogne et al. 2005; Krivitzky et al. 2009; Seminara et al. 2010).

We report on an infant with argininosuccinate synthetase (ASS) deficiency who, despite a catastrophic neonatal presentation of hyperammonemic coma, currently presents adequate cognitive and neurological development over an 8-year follow-up period.

Case Report

This male infant was referred from a distant Northern community hospital at 8 days of life for seizures and coma.

He is the second child of a healthy non-consanguineous French–Canadian couple. Pregnancy was uneventful. Spontaneous vaginal delivery occurred at term. Birth weight was 3.3 kg (25–50th centile), height 52.5 cm (75th centile), and head circumference 34 cm (10–25th centile). APGAR scores were 9, 8, and 8 at 1, 5, and 10 min, respectively. He was a quiet baby who fed well on formula until day 6 of life when, over a 7-h period, he became lethargic and stopped feeding. He was brought to the local emergency department where investigations were undertaken. He was pale and poorly responsive, but the rest of his physical examination was otherwise unremarkable. Vital signs were notable for hypothermia with a rectal body temperature of 34.2°C (N, 34.4–37.8°C). Initial laboratory findings were: blood glucose, 2.4 mmol/L (N, $\geq$2.2 mmol/L and $\leq$7.8 mmol/L); capillary pH, 7.54 (N, 7.37–7.43); pCO_2, 26 torr (N, 26–41 torr); bicarbonate, 21.8 mmol/L (N, 16–24 mmol/L). The following were normal: complete blood count, plasma electrolytes levels, calculated anion gap (10), urinalysis (no ketosis), cerebrospinal fluid glucose and protein levels. A complete septic workup was undertaken, and intravenous hydration with antibiotic therapy initiated pending culture results. Nine hours after admission, focal seizures were noted, which increased in frequency over the next 6 h. The infant was treated with phenobarbital and diazepam and transferred to a secondary care center. At the time of transfer, his vital signs were: pulse 136, respiratory rate 54, blood pressure 102 over 77 mmHg, and rectal body temperature within the lower normal limit at 35.4°C.

Upon arrival at the second hospital, he presented a sustained tonic–clonic seizure and progressively became bradypneic. He was intubated and received additional doses of phenobarbital and phenytoin. Laboratory investigations were repeated with the inclusion of a blood ammonia level, which revealed hyperammonemia at 505 μmol/L (N, <110 μmol/L). Other liver function tests were normal. As neither ammonia scavenging medications nor hemodialysis were available at this distant hospital, transfer was immediately arranged to a tertiary care center in a metropolitan area (5-h plane flight). Preparations for immediate drug treatment and hemodialysis upon arrival at the tertiary care center were made during the time of aerial transport.

The infant was admitted to the intensive care unit with a Glasgow score of 3, severe hypotonia and generalized seizures. Vitals signs recorded on admission were: heart rate 60, blood pressure 82 over 53 mmHg, and rectal body temperature of 33.9°C. Initial ammonia level was 1,058 μmol/L. Drug treatment was promptly initiated with intravenous administration over 90 min of a loading dose of 250 mg/kg of each of sodium benzoate, sodium phenylacetate and L-arginine, followed by maintenance continuous infusion of 250 mg/kg/day of each of the previous with the addition of 100 mg/kg/day of L-carnitine. A caloric intake of 130 kcal/kg/day was reached with both intravenous 10% dextrose and 20% intralipid (1 g/kg/day) infusions.

Hemodialysis was instigated 2 h after admission. Ammonia level had by then decreased to 993 μmol/L. Body temperature at the time of initiation of dialysis was recorded at 34.0°C and progressive warming with, first, a heating blanket, then a heated bed, was initiated. Within 6 h of combined hemodialysis and drug treatment, blood ammonia level had dropped to 103 μmol/L. Glasgow score increased to 6, while body temperature had reached 36.6°C at the end of hemodialysis. Over the next 6 h, Glasgow score progressively improved and seizures stopped. Enteral feeding with protein-free formula (Mead Johnson 80056 Protein Free diet powder) was initiated 12 h after admission to reach an initial enteral caloric intake of 100 kcal/kg/day. Natural proteins were subsequently reintroduced 24 h after admission, at an initial dose of 0.3 g/kg/day.

Blood amino acids chromatography performed upon arrival was notable for elevated glutamine levels at 1,957 μmol/L (N, 474–736 μmol/L), citrulline at 3,949 μmol/L (N, 14–32 μmol/L), with low ornithine levels of 19 μmol/L (N, 25–103 μmol/L), and arginine at 15 μmol/L (N, 43–120 μmol/L) (Lepage et al. 1997). Urine orotic acid was measured at 36.53 μmol/mmol of creatinine (N, <3.85 μmol/mmol of creatinine). This biochemical profile was compatible with a diagnosis of citrullinemia type I [MIM 215700, ASS deficiency (EC 6.3.4.5)]. Subsequent sequencing analysis of the ASS1 gene (MIM 603470) revealed an homozygous c.1168G>A (p.G390R) missense mutation (NM_000050.4).

Following full clinical recuperation, the infant was discharged 10 days after admission. His blood ammonia was then 39 μmol/L. At discharge, protein intake was restricted at 1.1 g/kg/day, while total caloric intake was maintained at 140 kcal/kg/day. Pharmaceutical treatment at discharge was: sodium benzoate 290 mg/kg/day and L-arginine 250 mg/kg/day both divided four times daily.

He was evaluated monthly for metabolic control and general health until his first birthday. Follow-up appointments were subsequently gradually spaced, as metabolic control and clinical status were persistently adequate. Pharmaceutical treatment with sodium benzoate and L-arginine was pursued at 250 mg/kg/day each, while protein intake has remained at about 1.0 g/kg/day. Caloric intake was maintained at about 140 kcal/kg/day until 15 months of age, after which it was progressively lowered to reach age-appropriate needs by 2 years of age. He never suffered another hyperammonemic episode and is in general good health. Strict compliance to therapeutic and dietary regimen was observed.

His growth and global development are progressing steadily (weight at the 25th centile, height at the 10th centile, head circumference at the 25th centile). Blood ammonia values have consistently remained below 70 μmol/L. Over this 8-year follow-up period, citrulline levels have remained at a median of 1,500 μmol/L.

Seizures never recurred and subsequent electroencephalograms were normal. Formal neurological assessments were performed every 2 months during the first year of life, then annually. Slight lower limb hypertonia and abnormal ocular fixation were noted at 2 months of age. Occupational therapy and physiotherapy treatments were promptly initiated. Brain magnetic resonance imaging was performed at 9 months of age and was reported within normal limits, without brain atrophy or other signs of previous brain edema. The only notable finding was an extremely subtle T2-weighted hypersignal in the white matter of the subinsular cortex. Cortical imaging changes in the insular region have been reported as specific findings in hyperammonemia (Takanashi et al. 2003; Bindu et al. 2009; U-King-Im et al. 2011). Subsequent detailed neuro-ophthalmologic evaluation led to a diagnosis of isolated congenital motor nystagmus, unrelated to a previous neurological insult. Following neurological evaluations did not uncover significant developmental delay over time. No other cerebral imaging was performed.

A complete neuropsychiatric evaluation was undertaken at the age of 4, before school entry. The following intelligence and developmental stage scales were used: WPPSI-III, Leiter-R, and Griffiths. Global intelligence scores (WPPSI-III) and nonverbal intelligence scores (Leiter-R) were both within average. Griffiths developmental stages scores were of 4 years and 5 months for a chronological age of 4 years and 1 month (high average score) for all developmental aspects evaluated (gross motor skills, personal and social, performance) except for language (average, with some vocabulary and fluency difficulties) and for fine motor skills, which were assessed as borderline low to low average. Regular follow-ups in occupational therapy and speech and language therapy were maintained to this day. At 8 years of age, he functions well within a regular schooling system, participates in several out-of-school activities, and is described as an active, curious, and engaging boy.

Discussion

ASS deficiency (MIM 215700) is one of the most prevalent UCD worldwide (Uchino et al. 1998; Summar et al. 2008; Laróvere et al. 2009; Ibarra-González et al. 2010). Complete enzymatic deficiency results in classic or type I citrullinemia (CTLN1), which is typically associated with citrulline elevations above 1,500 μmol/L. Most citrullinemia type I patients who suffered their first hyperammonemic episode as neonates are recounted to have died or to face long-term developmental or neurological disabilities (Maestri et al. 1995; Tokatli et al. 1998; Uchino et al. 1998; Bachmann 2003a; Tuchman et al. 2008; Krivitzky et al. 2009; Seminara et al. 2010). Only two cases of ASS-deficient patients are reported with good outcome 2 years following rescue from neonatal hyperammonemic coma (Walter et al. 1992; Vilaseca et al. 2001).

Suggested factors of poor long-term prognosis for survivors of hyperammonemic coma are elevated initial or peak hyperammonemia level and duration and depth of coma (Msall et al. 1984; Bachmann 2003a; Gropman and Batshaw 2004; Tuchman et al. 2008). Msall and colleagues in particular described a significant negative correlation between duration of neonatal hyperammonemic coma (but not with peak ammonia value) and subsequent IQ scores at 12 months. As well, age at onset of symptoms, residual enzymatic activity, compliance to dietary and drug regimen, and subsequent hyperammonemic episodes can each alter long-term outcome (Maestri et al. 1991; Saudubray et al. 1999; Nassogne et al. 2005; Tuchman et al. 2008; Seminara et al. 2010).

Peak ammonia levels for the two neonates described by Vilaseca et al. (2001) and Walter et al. (1992) were of 780 μmol/L and 2,500 μmol/L, respectively, but neither level nor duration of coma, dietary, or long-term management measures were precisely stated in those two reports. Although ammonia level at initial presentation is not available, the following is an estimate of our patient's clinical evolution timeline. By the time the child arrived at our center, it had been 31 h since onset of symptoms (lethargy, poor feeding) and 24 h since biochemical signs associated with hyperammonemia were documented (respiratory alkalosis). Clinical coma duration upon arrival at our center was estimated at about 7 h (from onset of tonic–clonic seizures associated with bradypnea). Thus, total duration of hyperammonemia (from onset of symptoms to clinical resolution of both symptoms and hyperammonemia) was estimated at nearly 39 h, while total duration of hyperammonemic coma was estimated at about 14 h. Peak ammonia value was documented at 1,058 μmol/L approximately 8 h after onset of coma. Therefore, our patient fits Msall and colleagues' model, where good outcome is correlated with short coma duration (below 24 h) but not with peak ammonia level.

Survival from hyperammonemic coma has dramatically improved with access to current treatment protocols (Enns et al. 2007). However, intellectual function data from a recent cohort study are sobering: for children aged 3–16 years who presented with neonatal-onset hyperammonemia, half had functional IQ values in the range of intellectual disability, with two-third of those in the moderate to profound disability range, ASS-deficient patients being the most severely affected (Seminara et al. 2010). Moreover, longer follow-ups tend to demonstrate progressive declines in IQ, motor skills deficits, or attention deficits in citrullinemia type I patients rescued from neonatal hyperammonemic coma, even in those who do not suffer subsequent hyperammonemic episodes (Maestri et al. 1995; Krivitzky et al. 2009; Seminara et al. 2010). Although liver transplantation is curative of the metabolic derangement and removes the necessity of strict dietary regimen observance and pharmaceutical treatment for UCD patients, neurologic outcome of these transplanted UCD patients is usually closely correlated to their pretransplantation neurologic status (Whitington et al. 1998; Saudubray et al. 1999; Nassogne et al. 2005; Stevenson et al. 2010). Moreover, hepatic transplantation is a complex surgical procedure with significant risks of peri- and postoperative complications, mortality, and organ rejection. Lifelong immunosuppressive therapy is required to maintain engraftment. Living donor grafts from asymptomatic family members is possible, but unrelated donors are few.

Hyperammonemia-induced brain toxicity is thought to occur through several mechanisms: conversion of ammonia to osmotically active glutamine resulting in cerebral edema, increased nitric oxide synthesis and oxidative stress, glutamine-mediated alteration in mitochondrial permeability and metabolism (Cagnon and Braissant 2007; Albrecht et al. 2010; Braissant 2010), and potentially altered neurotransmitter pathways (Enns 2008). Current therapeutic modalities aim to rapidly lower blood ammonia levels and thus reduce its detrimental effects on brain cells. Other currently investigational treatment options to prevent ammonia-induced brain injury include low-grade controlled systemic hypothermia, hypothesized to reduce brain energy demand, both in terms of glucose and oxygen metabolism, restore cerebral blood flow regulation, and decrease intestinal ammonia production (Whitelaw et al. 2001; Chatauret et al. 2002; Stravitz and Larsen 2009).

In our patient, mild hypothermia was neither therapeutically induced nor controlled, but may have played a role both in rapid response to ammonia-lowering therapeutic measures and in neuroprotection. Other factors that had a probable positive impact on our patient's immediate recovery and positive long-term outcome are: rapid institution of hemodialysis, rapid return to enteral feeding once coma had resolved, optimal caloric intake both during the acute episode and during long-term management, strict compliance to pharmaceutical and dietary regimen, and active prevention of subsequent hyperammonemic episodes.

Genotype–phenotype correlations as an aid to clinical management of ASS-deficient patients have been attempted. Wide phenotypic variability has been reported among individuals with ASS deficiency, even among members of the same family (Issa et al. 1988). The c.1168G>A (p.G390R) missense ASS1 variant identified in our patient is so far the single most common variant identified in early-onset citrullinemia type I (Kobayashi et al. 1990; Häberle et al. 2002; Gao et al. 2003; Engel et al. 2009; Laróvere et al. 2009). Homozygosity for this variant is usually associated with undetectable or severely reduced (less than 2%) residual enzymatic activity (Kobayashi et al. 1990; Vilaseca et al. 2001; Gao et al. 2003; Berning et al. 2008). Although associated with severe enzymatic deficiency and neonatal-onset hyperammonemia, this genotype does not systematically equate with poor long-term outcome, as demonstrated in our patient.

Newborn screening aims in part at instituting rapid therapeutic interventions for infants at risk of life-threatening disorders. Citrullinemia neonatal screening permits early treatment of newborns at risk and likely affects both survival and cognitive outcome. One caveat of citrullinemia newborn screening is that most patients with severe, early-onset forms usually present clinically before screening results are available (Bachmann 2003b; Wilcken et al. 2009). Moreover, newborn screening programs worldwide have now identified individuals with elevated citrullinemia (>3,000 μmol/L) who remain asymptomatic (Wilcken et al. 2009; Wilcken 2010). Debate on how these patients should be managed is still ongoing, although recent reports of asymptomatic adult citrullinemia patients experiencing their first hyperammonemic episode during pregnancy or the postpartum period do argue in favor of protein restriction even for "mild" cases.

Thus, several inherited and environmental factors play important roles in the initial presentation, therapeutic response, and long-term evolution of patients with ASS deficiency. Reliable clinical indicators of disease course and severity are currently lacking. There are inherent difficulties in comparing and interpreting the data of survivors of neonatal hyperammonemic episodes published over the past 20 years: lack of uniform initial clinical management across medical centers and lack of systematic documentation of clinical elements that could potentially modify treatment response or long-term evolution. Detailed prospective clinical registries as well as longitudinal interventional studies are essential to resolve this important issue, as an event-free course and optimal long-term quality of life are the goal to reach for these patients. The Urea Cycle Disorders Consortium, a network of 15 clinical and research centers providing care for UCD patients, has established a national research and clinical registry in the USA with collaborators in Canada and in Switzerland that will help in providing comprehensive data to improve clinical management and outcome for UCD patients.

Conclusion

Although the prognosis of neonatal hyperammonemic coma is traditionally considered as guarded, both in terms of survival and long-term neurological outcome, our case shows that recuperation from a severe neonatal hyperammonemic episode without long-term impairment is possible. The following factors were likely influential in the favorable evolution of our patient: rapid therapeutic intervention with appropriate ammonia-lowering measures and optimal caloric intake to reverse catabolism, fast lowering of blood ammonia during the acute episode, short hyperammonemic coma duration, strict long-term metabolic control with optimal compliance to pharmaceutical and dietary regimen, as well as aggressive prevention of subsequent hyperammonemic episodes. Registries and prospective studies in neonates treated with the current standardized clinical protocols will prove invaluable in assessing the precise impact of each of these elements in the long-term management of these high-risk patients.

Acknowledgments This chapter is dedicated to Yolande Lefèvre, registered nurse, who devoted her lifelong career to the care of children with metabolic disorders. We thank Dr. Jean-Claude Décarie for expert revision of the MRI imaging findings.

Synopsis

We describe the favorable long-term neurological outcome of a boy with argininosuccinate synthetase deficiency who presented with severe neonatal hyperammonemic coma.

References to Electronic Databases

Argininosuccinate synthetase 1; citrullinemia, classic: MIM 603470; 215700

Argininosuccinate synthetase: EC 6.3.4.5

ASS1 gene: geneID 445; NM_000050.4

Competing Interest Statement

Each author declares that the answer to all questions on the JIMD Reports competing interest form are "No," and therefore they have nothing to declare.

References

Albrecht J, Zielińska M, Norenberg MD (2010) Glutamine as a mediator of ammonia neurotoxicity: a critical appraisal. Biochem Pharmacol 80:1303–1308

Bachmann C (2003a) Outcome and survival of 88 patients with urea cycle disorders: a retrospective evaluation. Eur J Pediatr 162:410–416

Bachmann C (2003b) Long-term outcome of patients with urea cycle disorders and the question of neonatal screening. Eur J Pediatr 162: S29–S33

Berning C, Bieger I, Pauli S et al (2008) Investigation of citrullinemia type I variants by in vitro expression studies. Hum Mutat 29:1222–1227

Bindu PS, Sinha S, Taly AB et al (2009) Cranial MRI in acute hyperammonemic encephalopathy. Pediatr Neurol 41:139–142

Braissant O (2010) Current concepts in the pathogenesis of urea cycle disorders. Mol Genet Metab 100:S3–S12

Cagnon L, Braissant O (2007) Hyperammonemia-induced toxicity for the developing central nervous system. Brain Res Rev 56:183–197

Chatauret N, Rose C, Butterworth RF (2002) Mild hypothermia in the prevention of brain edema in acute liver failure: mechanisms and clinical prospects. Metab Brain Dis 17:445–451

Engel K, Höhne W, Häberle J (2009) Mutations and polymorphisms in the human argininosuccinate synthetase (ASS1) gene. Hum Mutat 30:300–307

Enns GM (2008) Neurologic damage and neurocognitive dysfunction in urea cycle disorders. Semin Pediatr Neurol 15:132–139

Enns GM, Berry SA, Berry GT et al (2007) Survival after treatment with phenylacetate and benzoate for urea-cycle disorders. N Engl J Med 356:2282–2292

Gao HZ, Kobayashi K, Tabata A et al (2003) Identification of 16 novel mutations in the argininosuccinate synthetase gene and genotype-phenotype correlation in 38 classical citrullinemia patients. Hum Mutat 22:24–34

Gropman AL, Batshaw ML (2004) Cognitive outcome in urea cycle disorders. Mol Genet Metab 81:S58–S62

Häberle J, Pauli S, Linnebank M et al (2002) Structure of the human argininosuccinate synthetase gene and an improved system for molecular diagnostics in patients with classical and mild citrullinemia. Hum Genet 110:327–333

Ibarra-González I, Fernández-Lainez C, Vela-Amieva M (2010) Clinical and biochemical characteristics of patients with urea cycle disorders in a developing country. Clin Biochem 43:461–466

Issa AR, Yadav G, Teebi AS (1988) Intrafamilial phenotypic variability in citrullinaemia: report of a family. J Inherit Metab Dis 11:306–307

Kobayashi K, Jackson MJ, Tick DB et al (1990) Heterogeneity of mutations in argininosuccinate synthetase causing human citrullinemia. J Biol Chem 265:11361–11367

Krivitzky L, Babikian T, Lee HS et al (2009) Intellectual, adaptive, and behavioral functioning in children with urea cycle disorders. Pediatr Res 66:96–101

Laróvere LE, Angaroni CJ, Antonozzi SL et al (2009) Citrullinemia type I, classical variant. Identification of ASS-p G390R (c.1168G>A) mutation in families of a limited geographic area of Argentina: a possible population cluster. Clin Biochem 42:1166–1168

Leonard JV, McKiernan PJ (2004) The role of liver transplantation in urea cycle disorders. Mol Genet Metab 81:S74–S78

Lepage N, McDonald N, Dallaire L et al (1997) Age-specific distribution of plasma amino acid concentrations in a healthy pediatric population. Clin Chem 43:2397–2402

Maestri NE, Hauser ER, Bartholomew D et al (1991) Prospective treatment of urea cycle disorders. J Pediatr 119:923–928

Maestri NE, Clissold DB, Brusilow SW (1995) Long-term survival of patients with argininosuccinate synthetase deficiency. J Pediatr 127:929–935

Meyburg J, Hoffmann GF (2010) Liver, liver cell and stem cell transplantation for the treatment of urea cycle defects. Mol Genet Metab 100:S77–S83

Msall M, Batshaw ML, Suss R et al (1984) Neurologic outcome in children with inborn errors of urea synthesis. Outcome of urea-cycle enzymopathies. N Engl J Med 310:1500–1505

Nassogne MC, Héron B, Touati G et al (2005) Urea cycle defects: management and outcome. J Inherit Metab Dis 28:407–414

Saudubray JM, Touati G, Delonlay P et al (1999) Liver transplantation in urea cycle disorders. Eur J Pediatr 158:S55–S59

Seminara J, Tuchman M, Krivitzky L et al (2010) Establishing a consortium for the study of rare diseases: the urea cycle disorders consortium. Mol Genet Metab 100:S97–S105

Stevenson T, Millan MT, Wayman K et al (2010) Long-term outcome following pediatric liver transplantation for metabolic disorders. Pediatr Transplant 14:268–275

Stravitz RT, Larsen FS (2009) Therapeutic hypothermia for acute liver failure. Crit Care Med 37:S258–S264

Summar ML, Dobbelaere D, Brusilow S et al (2008) Diagnosis, symptoms, frequency and mortality of 260 patients with urea cycle disorders from a 21-year, multicentre study of acute hyperammonaemic episodes. Acta Paediatr 97:1420–1425

Takanashi J, Barkovich AJ, Cheng SF et al (2003) Brain MR imaging in neonatal hyperammonemic encephalopathy resulting from proximal urea cycle disorders. AJNR Am J Neuroradiol 24(6):1184–1187

Tokatli A, Coşkun T, Ozalp I (1998) Citrullinemia. Clinical experience with 23 cases. Turk J Pediatr 40:185–193

Tuchman M, Lee B, Lichter-Konecki U et al (2008) Urea cycle disorders consortium of the rare diseases clinical research network. Cross-sectional multicenter study of patients with urea cycle disorders in the United States. Mol Genet Metab 94:397–402

Uchino T, Endo F, Matsuda I (1998) Neurodevelopmental outcome of long-term therapy of urea cycle disorders in Japan. J Inherit Metab Dis 21:S151–S159

U-King-Im JM, Yu E, Bartlett E et al (2011) Acute hyperammonemic encephalopathy in adults: imaging findings. AJNR Am J Neuroradiol 32:413–418

Vilaseca MA, Kobayashi K, Briones P et al (2001) Phenotype and genotype heterogeneity in mediterranean citrullinemia. Mol Genet Metab 74:396–398

Walter JH, Allen JT, Holton JB (1992) Arginosuccinate synthetase deficiency: good outcome despite severe neonatal hyperammonaemia. J Inherit Metab Dis 15:282–283

Whitelaw A, Bridges S, Leaf A et al (2001) Emergency treatment of neonatal hyperammonaemic coma with mild systemic hypothermia. Lancet 358:36–38

Whitington PF, Alonso EM, Boyle JT et al (1998) Liver transplantation for the treatment of urea cycle disorders. J Inherit Metab Dis 21:S112–S118

Wilcken B (2010) Expanded newborn screening: reducing harm, assessing benefit. J Inherit Metab Dis 33:S205–S210

Wilcken B, Haas M, Joy P et al (2009) Expanded newborn screening: outcome in screened and unscreened patients at age 6 years. Pediatrics 124:e241–e248

JIMD Reports
DOI: 10.1007/8904_2011_22

Received: 27 October 2010; Revised: 31 January 2011;
Accepted: 1 February 2011; Published online: 22 June
Communicated by: Ertan Mayatepek

Stroke and Stroke-Like Symptoms in Patients with Mutations in the POLG1 Gene

Waleed Brinjikji, Jerry W. Swanson, Carrie Zabel, Peter J. Dyck, Jennifer A. Tracy, and Ralitza H. Gavrilova

Abstract *Introduction/Methods* Mutations in *POLG1*, the gene encoding mitochondrial polymerase gamma (Polγ), have been associated with a number of well-characterized phenotypes. In this study, we report two cases of patients with biallelic *POLG1* mutations and stroke. We also performed a review of the literature and report on all clinical studies of patients with *POLG1* mutations in which stroke was described in the phenotype. For each patient, genotype and phenotype are reported.

Results Including our two patients, a total of 22 patients have been reported with *POLG1* mutations and stroke. The average age of onset of stroke in these patients was 9 years with a range of 1–23 years. In cases where localization was reported, the occipital lobes were the primary location of the infarct. Mutations in the linker–linker or linker–polymerase domains were the most frequent genotype observed. Seizures (16/22) and hepatic dysfunction/failure (8/22) were the most commonly reported symptoms in the stroke cohort.

Conclusion This article raises an underrecognized point that patients with *POLG1* mutations may suffer a cerebrovascular accident at a young age. The most common location of the infarction is in the occipital lobe. The presentation may be similar to MELAS and can be misdiagnosed as a migrainous stroke.

Keywords Mitochondrial · POLG1 · Stroke

W. Brinjikji
Mayo Medical School, Mayo Clinic, Rochester, MN, USA

J.W. Swanson, P.J. Dyck, and J.A. Tracy
Department of Neurology, Mayo Clinic, Rochester, MN, USA

C. Zabel, and R.H. Gavrilova (✉)
Department of Medical Genetics, Mayo Clinic, 200 1st Street SW, Rochester, MN 55905, USA
e-mail: gavrilova.ralitza@mayo.edu

Introduction

Mitochondrial DNA polymerase γ (Polγ) is the only known human mitochondrial DNA (mtDNA) polymerase. This enzyme plays an essential role in maintaining mtDNA integrity. Biallelic mutations in *POLG1* cause a deficiency of Polγ and lead to a variety of phenotypes. Bernard Alpers began to delineate the history of *POLG1*-related disorders in 1931, when he published the first case of diffuse progressive gray matter degeneration in children with severe epilepsy and cortical blindness (Alpers 1931). In 1999, Naviaux was the first to report POLG1 dysfunction and mtDNA depletion in a patient with Alpers syndrome (Naviaux et al. 1999). Since its discovery less than a decade ago, more than 150 mutations in *POLG1* have been associated with a spectrum of phenotypes including Progressive External Ophthalmoplegia (PEO), Alpers–Huttenlocher Syndrome (psychomotor retardation, intractable epilepsy, liver failure), and Ataxia Neuropathy Spectrum Disorders, among others (Blok et al. 2009; Milone and Massie 2010; Rahman and Hanna 2009; Van Goethem et al. 2001; Wong et al. 2008).

In this report, we describe two patients with Ataxia-Neuropathy (AN) and stroke due to *POLG1* mutations and provide a review of the literature.

Cases

Case 1

A 28-year-old woman presented to clinic for evaluation of progressive peripheral neuropathy symptoms of 8 years duration. Her peripheral neuropathy symptoms were primarily sensory and consisted of decreased sensation, paresthesias in her feet, and shooting pains in her legs. She was born full term by vaginal delivery with a birth weight of 7 lb and no perinatal complications. She had no significant childhood illnesses prior to the age of 7, when she was hospitalized for a week long episode of headache, fever, lethargy, and seizures. She was found to have bilateral occipital infarcts also associated with visual field deficits. Infectious workup was negative and cerebral angiography revealed no abnormalities. A diagnosis of MELAS (mitochondrial myopathy, encephalopathy, lactic acidosis, and stroke-like episodes) was suspected. However, sequencing of the entire mtDNA did not reveal any known or potentially pathogenic mutations known to cause MELAS or other mitochondrial disorder. Subsequent to this episode, bilateral hemianopsia, intractable partial seizures, and migraine headaches persisted. The patient began valproate therapy at age 11 and suffered hepatic encephalopathy which resolved after carnitine therapy. At age 21, she first noted a burning sensation in her feet and gait imbalance which had since progressed. Her family history was negative for others with similar complaints.

Plasma creatinine, lactate, ALT, and creatine kinase were normal. Spinal fluid analysis revealed an elevated total protein (122 mg/dl nl$\leq$60 mg/dl); there were no oligoclonal bands. Brain MRI demonstrated focal areas of T2 signal abnormality and volume loss in the occipital lobes bilaterally, left greater than right, consistent with old infarctions; there was no sign of recent ischemia (Fig. 1). EEG showed a mild degree of nonspecific generalized slowing, maximal over the left occipital head region. EMG of the left lower extremity revealed a length-dependent sensory peripheral neuropathy.

Gene sequencing of *POLG1* identified a c.1399G$\rightarrow$A (A467T) mutation on both alleles (e.g., the homozygous state) and confirmed a diagnosis of *POLG1*-Related Ataxia-Neuropathy Spectrum Disorder.

Case 2

A 39-year-old man presented for evaluation of progressive gait ataxia of 15 years duration; past medical history was also significant for migraine, right occipital infarction with residual left homonymous heminanopsia, seizure disorder, peripheral neuropathy, ophthalmoplegia, and myoclonus. The patient was healthy until age 18 when he suffered an acute right occipital infarction in the setting of an excruciating headache and a secondary generalized seizure. Four-vessel cerebral angiogram and spinal fluid analysis were normal. His initial diagnosis was migraine-related cerebral infarction. He was left with a residual left homonymous hemianopsia. An attempt to treat his seizures with valproate resulted in depressed mood and mild encephalopathy; he discontinued valproate therapy 6 weeks later. In his mid-1920s, he began to develop progressive gait unsteadiness, mild upper-extremity ataxia, and was found to have evidence of peripheral neuropathy. In his mid-1930s, he began to experience mild, but progressive, dysarthria and had occasional episodes of dysphagia. He had no significant family history of similar complaints.

Serum creatinine, lactate, ALT, and creatine kinase were normal. Brain MRI revealed findings consistent with an old infarction in the posteromedial right occipital lobe (Fig. 2). EMG findings were consistent with a mild length-dependent sensorimotor axonal peripheral neuropathy with no evidence for a myopathy. Echocardiography revealed no abnormalities.

POLG1 gene sequencing revealed homozygosity for the c.1399G$\rightarrow$A (A467T) mutation and the patient was diagnosed with *POLG1*-Related Ataxia-Neuropathy Spectrum Disorder. Following his diagnosis, he was placed on a mitochondrial cocktail consisting of alpha-lipoic acid, levocarnitine, and coenzyme Q10 and was advised to take L-arginine in light of his history of stroke. He is currently 40 years of age and his condition remains stable.

Discussion

POLG1 Mutations and Stroke

POLG1 mutations are associated with a broad range of clinical phenotypes including Alpers–Huttenlocher syndrome, PEO, and Ataxia-Neuropathy Spectrum Disorders; however, reports of cases involving stroke are rare (Blok et al. 2009; Milone and Massie 2010; Rahman and Hanna

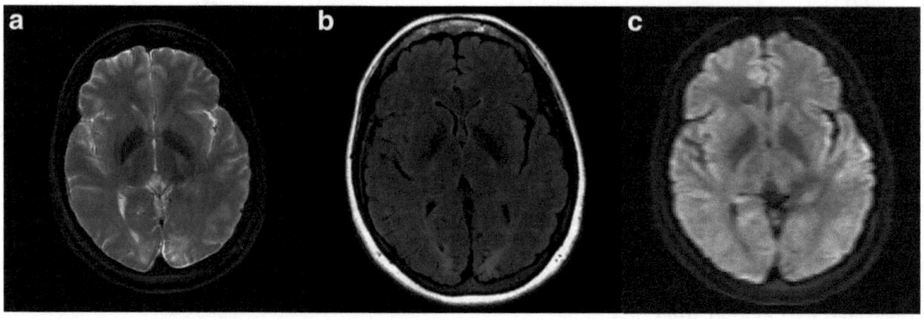

Fig. 1 (**a**, **b**) T2 and flair sequences demonstrate focal areas of T2 signal abnormality and volume loss in the left greater than right occipital lobes, consistent with prior infarcts. (**c**) Diffusion-weighted imaging is negative for recent ischemia

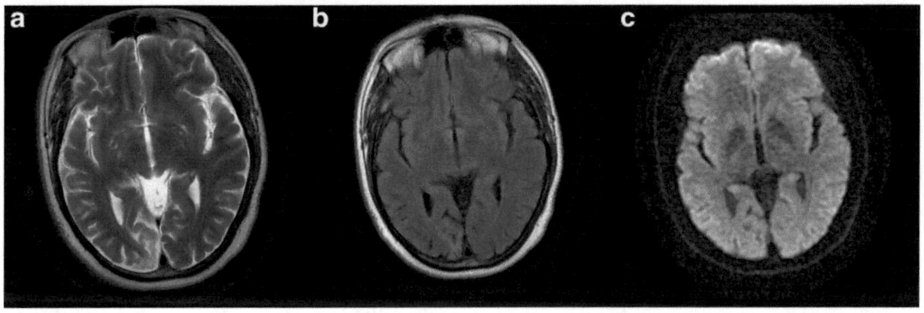

Fig. 2 (**a**, **b**) T2 and flair sequences demonstrate encephalomalacia involving the posteromedial right occipital lobe at the site of a prior infarct. (**c**) No recent sites of cortical infarction on diffusion imaging

2009; Stewart et al. 2009; Van Goethem et al. 2001; Wong et al. 2008). Including the above two cases, we found 22 reported cases of *POLG1* mutations associated with stroke and stroke-like symptoms in the literature (Table 1) (Blok et al. 2009; Deschauer et al. 2007; Horvath et al. 2006; Kollberg et al. 2006; Wong et al. 2008). Five of these 22 patients were clinically characterized to have Ataxia-Neuropathy Spectrum phenotype, 13 had a phenotype consistent with Alpers, and four patients were unclassified. Given the severity of the phenotypes observed among members of the stroke cohort, phenotypes of moderate to high severity might be more likely to be further complicated by stroke.

The region of *POLG1* between the exonuclease and polymerase has been termed the linker region (Fig. 3). Both of our patients were homozygous for the c.1399G→A (A467T) mutation located in the linker region of the *POLG1* gene. Linker mutations are the most common causes of *POLG1*-related mitochondrial disorders, among which the A467T substitution is the most frequently observed (Chan et al. 2005a; Luoma et al. 2005). In vitro studies of the effects of the A467T mutation suggest that it leads to severely reduced mtDNA polymerase activity (4% of wild type mtDNA polymerase activity) and disrupts binding of the POLG2 accessory subunit (Chan et al. 2005b; Tzoulis et al. 2006). The impaired accessory subunit interaction and reduced polymerase activity is the likely mechanism responsible for mtDNA depletion (Chan et al. 2005b).

Mutations of *POLG1* may also occur in the polymerase domain; these alterations affect mtDNA replication and can lead to reduced mtDNA copy number or to mtDNA deletions. Within the stroke cohort, all but one patient had at least one mutation in the linker region. Five patients were observed to have both mutations in the linker region, 11 patients were compound heterozygotes for mutations in both the linker and polymerase region, and one patient was a compound heterozygote for mutations in both the linker and exonuclease region (Fig. 3). Four patients had only one mutation detected, three of which were in the linker region and one in the polymerase domain of *POLG1*. These patients are suspected to have a second mutation, which is undetectable by current molecular genetic sequencing techniques.

The average age of onset of stroke or stroke-like symptoms in these patients was 9 years with a range of 1–23 years. Other commonly associated symptoms, in order of decreasing frequency, were: seizures (16/22), hepatic dysfunction/failure (8/22), myoclonus (6/22), peripheral neuropathy (3/22), migraine (3/22), and myopathy (3/22). There was no consistent constellation of symptoms associated with *POLG1* mutations and stroke among these patients.

Of the cases where the localization of the stroke was noted, 7/13 were located in the occipital region (Blok et al. 2009; Deschauer et al. 2007). MRI findings were available in nine patients. In five of nine patients, increased T2 signal was present in the occipital lobes. These findings were consistent with old infarction. Two patients had increased T2 signal in the basal ganglia, one in the basal ganglia and right frontal lobe and the other in the right frontal lobe without abnormalities in the occipital lobes. Laboratory findings were available in 15 patients. The most common laboratory abnormalities included elevated serum lactate (9/15) and elevated liver transaminases (7/15).

Similarity to Stroke in MELAS

The first patient discussed (Case 1) was initially thought to have MELAS because of the presence of cerebral infarction. MELAS is a well-characterized syndrome associated with stroke-like episodes due to mutations in mtDNA. Stroke-like episodes in MELAS are often accompanied by a severe migraine-like headache and/or seizures. Many patients also suffer from lactic acidosis or encephalopathy. Cerebral infarctions in MELAS are generally localized in the parietooccipital regions and are not in any particular vascular distribution (Rahman and Hanna 2009).

Patients with MELAS have been found to have impaired endothelial function and vasoconstriction when compared to controls. Therefore, one of the theories for the etiology of stroke in MELAS is impaired vasodilation and a deficiency of nitric oxide (NO) (Koga et al. 2007). Some have studied the benefits of administering L-arginine, a precursor to the vasodilator NO in patients with MELAS stroke-like episodes

Table 1 Patients with POLG1 mutation and stroke. *ANS* ataxia neuropathy spectrum, *MELAS* mitchondrial myopathy, encephalopathy, lactic acidosis, stroke, *U* Unknown, *PEO* progressive external ophthalmoplegia

Patient	Study	Family history	Sex	Age of onset	Phenotype: symptoms	Lab findings	MRI findings	Allele 1 Domain	Allele 2 Domain
1	Current study	None	M	18	ANS: occipital stroke, visual field deficit, peripheral neuropathy, migraine, epilepsy, ataxia, myoclonus	Elevated CSF protein	T2 signal abnormality and volume loss in left > right occipital lobes consistent with prior infarcts	1399G→A (A467T) Linker	1399G→A (A467T) Linker
2	Current study	None	F	7	ANS: occipital stroke, visual field deficit, peripheral neuropathy, epilepsy, migraine, Valproate-induced hepatic necrosis	No abnormal labs	T2 hyperintensity in the posteromedial right occipital lobe, consistent with prior infarct	1399G→A (A467T) Linker	1399G→A (A467T) Linker
3	Horvath et al., Brain (2006)	None	M	7	Alpers: stroke-like episode, seizures	–	–	1880G→A (R627Q) Linker	3287G→A (R1096H) Polymerase
4	Horvath et al., Brain (2006)	None	M	4	Alpers: stroke-like episode, myoclonus	–	–	1399G→A (A467T) Linker	2740A→C (T914P) Polymerase
5	Horvath et al., Brain (2006)	Consanguineous	M	8	Alpers: stroke-like episode	–	–	1399G→A (A467T) Linker	1399G→A (A467T) Linker
6	Wong et al., Hum Mutat (2008)	None	M	1	Alpers: stroke, hypotonia	Elevated serum lactate and liver transaminases	–	1399G→A (A467T) Linker	2157+5_+6GC→AG Linker
7	Wong et al., Hum Mutat (2008)	None	F	2	Alpers: stroke	Elevated serum lactate and liver transaminases	–	1399G→A (A467T) Linker	2542G→A (G848S) Polymerase
8	Wong et al., Hum Mutat (2008)	None	F	9	Alpers: stroke, ataxia, exercise intolerance	Elevated liver transaminases	–	1399G→A (A467T) Linker	2542G→A (G848S) Polymerase
9	Wong et al., Hum Mutat (2008)	None	F	15	ANS: stroke, chorea, ataxia, ptosis, retinitis pigmentosa, liver transplant	Elevated serum lactate and liver transaminases	–	2554C→T (R852C) Polymerase	32G→A (G11D) Exonuclease 1880G→C (R627Q) Linker
10	Wong et al., Hum Mutat (2008)	None	F	16	U: occipital stroke, short stature, Leigh-like disease, neuropathy, dystonia, chorea, diabetes, myoclonus	No reported abnormal labs	–	1550G→T (G517V) Linker	–

11	Blok et al., J Med Genet (2009)	None	F	19	U: MELAS-like features, occipital epilepsy	–	–	1399G→A (A467T) Linker	2740A→C (T914P) Polymerase
12	Blok et al., J Med Genet (2009)	None	F	1	Alpers: epilepsy, deafness, retinitis pigmentosa, hemiparesis, liver failure	–	–	1399G→A (A467T) Linker	2772_2773 delG (K925Rfs42X) Polymerase
13	Blok et al., J Med Genet (2009)	None	F	1	Alpers: occipital strokes, growth retardation, focal epilepsy, liver failure, died of heart failure	–	–	1399G→A (A467T) Linker	2869G→C (A957P) Polymerase
14	Deschauer et al., Neurology (2007)	None	M	23	ANS: occipital stroke-like episode, headache, seizures, peripheral neuropathy, myopathy	Mild increase in CSF protein, elevated serum lactate	Increased T2 signaling from right occipital cortex sparing deeper white matter	1880G→A (R627Q) Linker	2542G→A (G848S) Polymerase
15	Kollberg et al., J Neuropathol Exp Neurol (2006)	Maternal grandmother: tremor and hepatopathy; Brother: tremor	F	1	Alpers: stroke-like episodes, epilepsy, ptosis, external ophthalmoplegia, valproate-induced liver failure, died at 4 years of multiple organ failure	Elevated serum and CSF lactate, elevated liver transaminase	Atrophy of both occipital lobes, high signal intensity in occipital lobe white matter	1399G→A (A467T) Linker	1720C→T (R574W) Linker
16	Kollberg et al., J Neuropathol Exp Neurol (2006)	Sister: POLG1 mutation and Alper Syndrome	M	11	ANS: migrainous headaches, ataxia, myoclonus, seizures, L hemiparesis, PEO, peripheral neuropathy	No reported abnormal labs	Increased T2 signaling from right motor cortex	2243G→C (W748S) 3428A→G Linker	695G→A (R232H) Exonuclease
17	Kollberg etal., J Neuropathol Exp Neurol (2006)	None	M	1	Alpers: transient L hemiparesis, myoclonus, seizures, died of liver failure at 13 months	Elevated serum lactate, liver transaminases and acylcarnitine fraction. Mitochondrial complex 1 deficiency	–	3488T→G (M1163R) Polymerase	–
18	Stewart et al., J Med Genet (2009)	None	N/A	1	Alpers: stroke-like episodes	–	–	32G→A (G11D) 2554C→T (R852C) Polymerase	2243G→C (W748S) Linker

(continued)

Table 1 (continued)

Patient	Study	Family history	Sex	Age of onset	Phenotype: symptoms	Lab findings	MRI findings	Allele 1 Domain	Allele 2 Domain
19	Montine et al., Clin Neuropathol (1995)	None	M	17	Alpers: severe headaches, multiple stroke-like episodes with visual deficits, seizures, hemorrhagic pancreatitis	Elevated serum lipase	Bilateral occipital lobe lesions with increased T2-weighted signal, non-enhancing	[a]	[a]
20	Hopkins et al., J Child Neurol (2010)	Grandmother: ataxia, bilateral basal ganglia infarcts, posterior spinal column degeneration. Identical twin suffered same symptoms	F	16	U: diabetes, basal ganglia strokes, choreiform movements, psychosis and epilepsy	Elevated serum lactate. Chronic myopathic changes	Increased T1 and decreased T2 in bilateral basal ganglia and thalami	1550G→T(G517V) Linker	–
21	Hopkins et al., J Child Neurol (2010)	Grandmother: ataxia, bilateral basal ganglia infarcts, posterior spinal column degeneration. Identical twin suffered same symptoms	F	17	U: diabetes, strokes, psychosis and epilepsy	Elevated serum lactate	Bilateral basal ganglia infarctions	1550G→T (G517V) Linker	–
22	Naviaux et al., Ann Neurol (2004)	Brother: alpers	M	1	Alpers: encephalopathy, ataxia, thalamic infarct	Elevated serum lactate and liver transaminases	CT: thalamic infarct	1399G→A (A467T) Linker	2899G→T (E873X) Polymerase

[a]No detected point mutations in mitochondrial DNA at positions 3243 or 3271 as reported in patients with MELAS

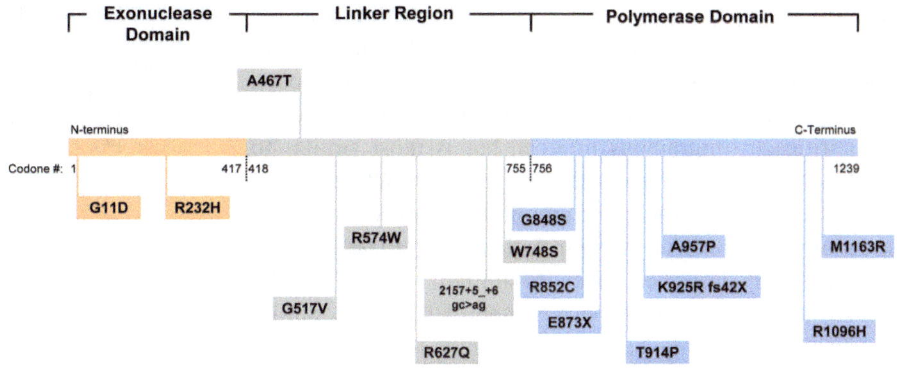

Fig. 3 Schematic representation of the POLG1 gene: exonuclease domain (codon 1–417), linker domain (codon 418–755), and polymerase domain (codon 756–1239)

(Koga et al. 2002, 2005, 2007). It has been found that intravenously administered L-arginine improved acute stroke-like symptoms in MELAS patients (Koga et al. 2005). Furthermore, long-term supplementation of oral L-arginine has been shown to improve endothelial dysfunction and potentially play a role in the prevention of stroke-like episodes in MELAS (Koga et al. 2006).

Similar to the MELAS stroke-like episodes, many patients with the *POLG1* mutation and stroke have reported migrainous headache, seizures, and cortical blindness in association with the stroke. Given the similarity of the stroke-like episodes in MELAS when compared to those with *POLG1* mutations and stroke, it is certainly warranted to investigate the role of endothelial dysfunction in their stroke-like episodes. Currently, no treatment recommendations exist for prevention of stroke in patients with *POLG1* mutations.

Similarity to Migrainous Infarction

As mentioned previously, Case 2 was initially diagnosed with a migrainous infarction. Migraine with aura is a well-established risk factor for ischemic stroke (Spector et al. 2010). According to the International Headache Society, migrainous infarction is defined as one or more migrainous aura symptoms associated with an ischemic brain lesion in the appropriate territory as demonstrated by neuroimaging. The diagnostic criteria include: (1) The attack in a patient with migraine with aura is typical of previous attacks, except that one or more aura symptoms persist for >60 min; (2) Neuroimaging demonstrates ischemic infarction in a relevant area; and (3) The infarction is not attributed to another disorder. Patients with migrainous infarction generally have neurological deficits in the vascular territory of the aura and have an associated lesion present on neuroimaging.

A number of potential mechanisms have been proposed for the migrainous stroke. These hypotheses include vasospasm, endothelial dysfunction, activation of the thrombotic cascade, increased platelet activation, and medications (namely triptans and ergotamine) (Del Zotto et al. 2008). Patients who have other stroke risk factors in the setting of migraine are also generally predisposed to an increased risk of stroke. Thus, a number of recommendations are in place to help prevent stroke in patients with migraine including treatment of modifiable risk factors (hypertension, diabetes, smoking, and hypercholesterolemia), avoiding estrogen containing oral contraceptives and medications with vasoconstrictive action (triptans), and migraine prophylaxis (Del Zotto et al. 2008).

When comparing the characteristics of stroke in patients with *POLG1* mutations and those of the migrainous stroke, there exist a number of similarities, but also a number of important differences. The two are similar in that the stroke can occur in the setting of a migraine or migraine-like episode. Patients with stroke and *POLG1* mutations have reported aura during their stroke episodes. However, migrainous strokes are generally not associated with seizure activity, lactic acidosis, or other associated symptoms. In addition, migrainous strokes typically occur in the vascular distribution of the aura and occur in patients older than those who possess *POLG1* mutations. If a patient presents with a migrainous stroke, it is important to follow this patient long-term to assure the lack of development of symptoms that might warrant further investigations for a mitochondrial cytopathy.

Conclusion

We describe 2 new cases and review a total of 22 cases of patients with stroke in the setting of disease-causing *POLG1* mutations. In general, these strokes occur in children under the age of 10, but can present at a later age. In cases where localization is reported, the occipital lobes are the primary location of the infarct. Linker–linker or linker–polymerase domain mutations were most commonly identified in the described patients with stroke in the setting of a more severe phenotype, such as Ataxia-Neuropathy Spectrum or Alpers. Stroke in the setting of *POLG1* mutations has a number of similarities to migrainous infarcts, but is most similar to stroke-like episodes in the MELAS syndrome. While neither a mechanism for stroke, nor any effective means of treatment or prophylaxis of stroke in patients with *POLG1* mutations has been described, given its similarity to the MELAS stroke, it is possible that these two entities share a similar mechanism. Given the benefit of L-arginine for patients with MELAS, it would be reasonable to study its effect in *POLG1*-related mitochondrial disorders.

Take-Home Statement

This article raises an important and underrecognized clinical point that stroke can occur in patients with *POLG1* mutations.

Mr. Waleed Brinjikji was involved in conception and design, or analysis and interpretation of data, and drafting the article or revising it critically for important intellectual content.

Dr. Jerry Swanson was involved in conception and design, or analysis and interpretation of data, and drafting the article or revising it critically for important intellectual content.

Ms. Carrie Zabel was involved in conception and design, or analysis and interpretation of data, and drafting the article or revising it critically for important intellectual content.

Dr. Peter Dyck was involved in drafting the article or revising it critically for important intellectual content.

Dr. Jennifer Tracy was involved in drafting the article or revising it critically for important intellectual content.

Dr. Ralitza Gavrilova was involved in conception and design, or analysis and interpretation of data, and drafting the article or revising it critically for important intellectual content. She is the GUARANTOR for the article.

The authors have no conflicts of interest to declare.

No ethics committee approval was necessary.

No funding was needed.

According to Mayo Clinic policy, patients give consent when they complete initial paperwork for their first visit. At that time, they are asked whether they have objections to use of information/images for publication purposes. In this case, no objections were stated.

References

Alpers BJ (1931) Diffuse progressive degeneration of the grey matter of the cerebrum. Arch Neurol Psychiatry 25:469–505

Blok MJ, van den Bosch BJ, Jongen E et al (2009) The unfolding clinical spectrum of POLG mutations. J Med Genet 46:776–785

Chan SS, Longley MJ, Copeland WC (2005a) The common A467T mutation in the human mitochondrial DNA polymerase (POLG) compromises catalytic efficiency and interaction with the accessory subunit. J Biol Chem 280:31341–31346

Chan SS, Longley MJ, Naviaux RK, Copeland WC (2005b) Mono-allelic POLG expression resulting from missense-mediated decay and alternative splicing in a patient with Alpers syndrome. DNA Repair 4:1381–1389

Del Zotto E, Pezzini A, Giossi A, Volonghi I, Padovani A (2008) Migraine and ischemic stroke: a debated question. J Cereb Blood Flow Metab 28:1399–1421

Deschauer M, Tennant S, Rokicka A et al (2007) MELAS associated with mutations in the POLG1 gene. Neurology 68:1741–1742

Hopkins SE, Somoza A, Gilbert DL (2010) Rare autosomal dominant POLG1 mutation in a family with metabolic strokes, posterior column spinal degeneration, and multi-endocrine disease. J Child Neurol 25:752–756

Horvath R, Hudson G, Ferrari G et al (2006) Phenotypic spectrum associated with mutations of the mitochondrial polymerase gamma gene. Brain 129:1674–1684

Koga Y, Ishibashi M, Ueki I et al (2002) Effects of L-arginine on the acute phase of strokes in three patients with MELAS. Neurology 58:827–828

Koga Y, Akita Y, Nishioka J et al (2005) L-arginine improves the symptoms of strokelike episodes in MELAS. Neurology 64:710–712

Koga Y, Akita Y, Junko N et al (2006) Endothelial dysfunction in MELAS improved by l-arginine supplementation. Neurology 66:1766–1769

Koga Y, Akita Y, Nishioka J et al (2007) MELAS and L-arginine therapy. Mitochondrion 7:133–139

Kollberg G, Moslemi AR, Darin N et al (2006) POLG1 mutations associated with progressive encephalopathy in childhood. J Neuropathol Exp Neurol 65:758–768

Luoma PT, Luo N, Löscher WN et al (2005) Functional defects due to spacer-region mutations of human mitochondrial DNA polymerase in a family with an ataxia-myopathy syndrome. Hum Mol Genet 14:1907–1920

Milone M, Massie R (2010) Polymerase gamma 1 mutations: clinical correlations. Neurologist 16:84–91

Montine TJ, Powers JM, Vogel FS, Radtke RA (1995) Alpers' syndrome presenting with seizures and multiple stroke-like episodes in a 17-year-old male. Clin Neuropathol 14:322–326

Naviaux RK, Nyhan WL, Barshop BA, Poulton J, Markusic D, Karpinski NC, Haas RH (1999) Mitochondrial DNA Polymerase gamma Deficiency and mtDNA Depletion in a Child with Alpers' Syndrome. Ann Neurol 45:54–58

Naviaux RK, Nguyen KV (2004) POLG mutations associated with Alpers' syndrome and mitochondrial DNA depletion. Ann Neurol 55:706–712

Rahman S, Hanna MG (2009) Diagnosis and therapy in neuromuscular disorders: diagnosis and new treatments in mitochondrial diseases. J Neurol Neurosurg Psychiatry 80:943–953

Spector JT, Kahn SR, Jones MR, Jayakumar M, Dalal D, Nazarian S (2010) Migraine headache and ischemic stroke risk: an updated meta-analysis. Am J Med 123:612–624

Stewart JD, Tennant S, Powell H et al (2009) Novel POLG1 mutations associated with neuromuscular and liver phenotypes in adults and children. J Med Genet 46:209–214

Tzoulis C, Engelsen BA, Telstad W et al (2006) The spectrum of clinical disease caused by the A467T and W748S POLG mutations: a study of 26 cases. Brain 14:1685–1692

Van Goethem G, Dermaut B, Lofgren A, Martin JJ, Van Broeckhoven C (2001) Mutation of POLG is associated with progressive external ophthalmoplegia characterized by mtDNA deletions. Nat Genet 28:211–212

Wong LJ, Naviaux RK, Brunetti-Pierri N et al (2008) Molecular and clinical genetics of mitochondrial diseases due to POLG mutations. Hum Mutat 29:E150–E172

JIMD Reports
DOI: 10.1007/8904_2011_23

Received: 3 September 2010; Revised: 20 December 2010;
Accepted: 27 January 2011; Published online: 22 June
Communicated by: Alberto Burlina

Perioperative Management of Hemostasis for Surgery of Benign Hepatic Adenomas in Patients with Glycogen Storage Disease Type Ia

Alix Mollet-Boudjemline, Aurélie Hubert-Buron, Catherine Boyer-Neumann, Roxana Aldea, Dominique Franco, Pascale Trioche-Eberschweiller, Anne-Elisabeth Mas, Mylène Mabille, Philippe Labrune, and Vincent Gajdos

Abstract The development of hepatocellular adenomas in the liver of patients with glycogen storage disease type I is a well-known complication of the disease. Surgical procedures and perioperative managements described so far have reported persistent and important morbidity. We report here a series of six patients (three males and three females) who underwent hepatic resection, and we propose a new hemostatic management protocol comprising glucose infusion, corticosteroids, desmopressin, and antifibrinolytic drugs, used to prevent efficaciously hepatic hemorrhage due to glycogen storage disease (GSD) platelet dysfunction.

Keywords GSD glycogen storage disease · Hemostasis · Hepatectomy · Platelet dysfunction

A. Mollet-Boudjemline (✉), A. Hubert-Buron,
P. Trioche-Eberschweiller, and V. Gajdos
APHP, Centre de Référence Maladies Héréditaires du Métabolisme Hépatique, Service de Pédiatrie, Hôpital Antoine Béclère, 157 Rue de la Porte de Trivaux, 92141 Clamart cedex, France
e-mail: philippe.labrune@abc.aphp.fr

C. Boyer-Neumann
APHP, Laboratoire d'hématologie biologique, Hôpital Antoine Béclère, Clamart cedex, France
and
UFR Kremlin Bicêtre, Univ-Paris Sud, Le Kremlin-Bicêtre, France

R. Aldea
APHP, Service d'anesthésie, Hôpital Antoine Béclère, Clamart cedex, France

D. Franco
APHP, Service de Chirurgie, Hôpital Antoine Béclère, Clamart cedex, France
and
UFR Kremlin Bicêtre, Univ-Paris Sud, Le Kremlin-Bicêtre, France

A.-E.Mas
APHP, Service d'anatomie pathologique, Hôpital Antoine Béclère, Clamart cedex, France

M. Mabille
APHP, Service de radiologie, Hôpital Antoine Béclère, Clamart cedex, France

P. Labrune
APHP, Centre de Référence Maladies Héréditaires du Métabolisme Hépatique, Service de Pédiatrie, Hôpital Antoine Béclère, 157 Rue de la Porte de Trivaux, 92141 Clamart cedex, France
and
UFR Kremlin Bicêtre, Univ-Paris Sud, Le Kremlin-Bicêtre, France
and
INSERM U 948, CHU Nantes, Nantes, France

Abbreviations

αFP	Alpha fetoprotein
BT	Bleeding time
CEA	Carcinoembryonic antigen
DDAVP	Desmopressin, deamino 8d arginine vasopressin
GDF	Gastric drip feeding
GOT	Glutamate oxaloacetic transaminase
GPT	Glutamate pyruvate transaminase
GSD	Glycogen storage disease
PFA	Platelet function aggregation
RBC	Red blood cell

Introduction

Glycogen Storage Disease Ia (GSD1a-Von Gierke's disease – OMIM 23220) is an autosomal recessive inborn error of carbohydrate metabolism caused by deficient activity of glucose-6-phosphatase (EC 3.1.3.9; P35575), an enzyme implied in gluconeogenesis and glycogenolysis. It leads to glycogen accumulation in liver, kidney, and intestinal mucosa. The clinical manifestations include steatotic hepatomegaly, growth difficulties and delayed puberty, hypoglycemia for short fasting time associated with lactic acidosis, hypertriglyceridemia, hyperuricemia (Rake et al. 2002a), and prolonged bleeding time (BT) due to impaired platelet function (Ambruso et al. 1985; Gilchrist et al. 1968; Hutton et al. 1976; Kao et al. 1980; Rake et al. 2002a). This last complication leads to potential hemorrhage during surgical procedures and to increased morbidity. The platelet dysfunction has been described secondary to a large excess of stored glycogen in the platelet, leading to decreased platelet adhesiveness and impaired platelet factor3 availability (Sahud 1972), due to chronic hypoglycemia (Hutton et al. 1976), or related to the presence of an inhibitor in GSD patients (Kao et al. 1980). Previous guidelines suggested the use of glucose infusion to stabilize bleeding diathesis (Rake et al. 2002b) and DDAVP has been used during bleeding complications in GSD patients (Marti et al. 1986). We report a series of six consecutive GSD Ia patients who underwent partial hepatectomy for complicated hepatic adenomas, one with a protocol stabilizing glucose homeostasis and lactic acidosis, and the other five with a personalized metabolic and hemostatic protocol, comprising desmopressin for four, to prevent hemorrhage. The current perioperative protocol is provided and discussed. We will compare clinical outcomes, hemostatic difficulties, and perioperative consequences before and after this management.

Case Reports

We report the cases of six patients with GSD type Ia who underwent partial hepatectomy for voluminous or complicated hepatic masses over the last 13 years. Specific clinical characteristics for each patient are presented in Table 1. Clinical diagnosis and initiation of treatment were achieved at median age of 21 months (range: 5–60 months). All patients had annual follow-up with a specialist in hepatic metabolism and received controlled treatment for GSDIa that included frequent meals, uncooked cornstarch alone or coupled with nocturnal gastric drip feeding of glucose. As proposed in our medical center, all patients had strict lactose restriction up to age five and moderate lactose restriction (<10 g/day) later on. They also were on fructose restriction limited to two fruit servings per day without restriction in vegetables. Through teenage and adulthood, compliance to treatment was poor for four patients. Two of the male patients used androgens as auto medication for growth purposes, or in a bodybuilding program. The third male patient was on growth hormone treatment during 24 months (from 15 to 17 years of age), initiated because of retardation to thrive with height <−4DS and pubertal delay (Tanner stage 1) with no height gain and no increased growth velocity. The three female patients were on progestative pills as contraception.

Hepatic masses were first detected at ages ranging from 13 to 16.25 years (mean 14.46 years). Ages of patients at the time of surgery ranged from 18 to 34 years (mean 24.17 years). Duration between the diagnosis of hepatic masses and surgical procedure ranged from 3.5 to 20.16 years (mean 9.71 years). Principal clinical characteristics related to surgery are presented in Table 2. Cancer biomarkers, αfetoprotein and CEA remained in normal range for all patients during follow-up. Histological analysis confirmed

Table 1 Clinical characteristics of the six patients

Patient	1	2	3	4	5	6
Gender	M	F	F	F	M	M
Age at diagnosis of GSD I (months)	8	10	5	7	60	36
Mode of treatment	DC	None	DC NEGI	DC NEGI	DC NEGI	DC NC
Treatment compliance (admitted)	+	−	−	−	−	+
Lipids concentration year of surgery (TG in g/l)	10.13	7.2	6.76	3.22	>30	20.4
Glucose nadir on cycle (mmol/l)	3.3	2.6	1.8	2.0	2.9	3.4
Mean lactate on glycemic cycle year of surgery (mmol/l)	5.55 (max 6.7)	5.67 (max 6.58)	7.6 (max 10)	5.9 (max 8.88)	5.06 (max 8.12)	4.92 (max 8.26)
Use of substitutive hormone therapy	Yes (for growth purposes)	Progestative pill	Progestative pill	Progestative pill	Androgens?	NoneGrowth hormone: 2 years, max 2 mg/day 6 days/7
Associated complications	Proteinuria hypertension osteoporosishTG> 35 g/l	Epileptic status (related to recurrent hypoglycaemia in infancy?)	Chronic anemia <8 g/dl resistant to iron supplements	Anorexia	Calcic lithiasis hTG> 20 g/l	Renal insufficiency and Moyamoya disease with residual left hemiparesia

DC daily cornstarch, *NC* nocturnal cornstarch, *NEGI* nocturnal enteral glucose infusion, *hTG* hypertriglyceridemia

Table 2 Clinical and surgical data of the six patients

Patient	1	2	3	4	5	6
Age at first diagnosis of adenoma (years)	13.84	16	13	13.16	16.25	14.5
Size of major adenoma at first diagnosis of tumor (cm)	2.1	3.65	2.0	0.8	3.4	2.4
Age at surgery (years)	34	23	24	26	20	18
Weight at surgery (kg/SD)	51/−2	62/+1.2	68/+2.7	52.0/0	87/3.53	38/−3.8
Height at surgery (cm/SD)	160/−2.2	162/−0.2	163/0	154/−2	175/0	148/−3.8
Size of major adenoma (cm)[a]	18	8	14	11	10.5	11
Number of adenomas[a]						
<2 cm	0	10	3	1	1	0
2–5 cm	1	5	4	2	4	0
5–10 cm	0	5	1	2	0	1
>10 cm	2	0	1	1	1	1
Complications of adenomas[a]	Asthenia, chronic abdominal pain, venous compression	Acute abdominal pain (sudden intra-tumoral hemorrhage)	Recurrent abdominal pains with central necrosis	Acute abdominal pain (sudden necrosis)	None	None
Indication for surgery	Complication	Complication	Size of adenoma +chronic pain	Complication	Size and rapid evolution of adenoma	Size and rapid evolution of adenoma
Surgical procedure	Right hepatectomy+ slight extension in segment IV	Right hepatectomy +atypical resection segments III and IV	Left hepatectomy	Right hepatectomy	Right hepatictomy+ atypical resection of segment IV+ RFA on segment II	Right hepatectomy
Estimated operative blood loss	>3,000 ml	700	<500 ml	<500 ml	<500 ml	900
Macromolecular filling	Yes	No	No	No	No	No
Peri-operative transfusion <72 h	7 RBC units+ 2 PU	No	No	No	No	6 platelet units+1 RBC unit
Nadir glucose (mmol/l=g/l)	3.0=0.54	6.8=1.2	4.7=0.84	4.8=0.86	3.7=0.67	3.8=0.69
	20.2	10.12	11.62	10.49	8.80	4.2

(continued)

Table 2 (continued)

Patient	1	2	3	4	5	6
Peak post-operative lactic acid (mmol/l)						
Other complications	None	None	None	Gazous emboly and peritonitis	Lymphan-gitis and entero-coccal infection of peri-hepatic collection	None except chronic denutrition
Hospital stay after surgery: intensive care+ standard hospitalization (days)	6+12	1+7	2+7	5+17	8+0	2+10
Evolution	Three small nodules (< p;2-cm) appearing 9 years after surgery, stable	No data (refuses assessment)	Rapid growth	Stable left in site adenomas	No residual adenoma at 1 year	No residual adenoma at 1 month

[a]At time of surgery
RBC red blood cell, *PU* platelet unit

the diagnosis of benign adenomas. For the last patient, adenomas were present before initiation of growth hormone and evolved rapidly (maximal diameter 2.4–14.5 cm in 3 years).

Case 1

The first patient was a 34-year-old male at the time of surgery with two hepatic adenomas diagnosed at 13.8 years. The voluminous masses, in the right liver measuring 18 × 12 cm in the junction between segments V and VI and 11 × 11 cm between segments VII and VIII, compressed the right portal branch with secondary alternate flow and led to frequent abdominal pain, anorexia, and asthenia (Fig. 1). Hemostatic delay (described in Table 3) was treated according to ESGSD I guidelines (Rake et al. 2002a) with 10% glucose infusion and hydrocortisone injections followed in postsurgical time by oral prednisone decreased in 12 days. Right hepatectomy, slightly extended to segment IV, was complicated by important hemorrhage (continuous bleeding) and hemodynamic instability necessitating numerous macromolecular filling, treatment by ephedrine and a blood transfusion of 5 units of RBC. Postoperatively, the patient had metabolic decompensation with liver insufficiency (Quick time 50% – Hb 7.7 g/dl – GOT 2,743U/L – GPT 1,443U/L), major lactic acidosis with hypoglycemia (pH 7.0; lactic acid 20.2 mmol/l). Major glucose and bicarbonate infusions and mechanic ventilation were necessary during 24 h. Normalization of acidosis markers was obtained in 2 days postoperatively. Thirteen years later, the patient is clinically well. We noted the apparition in the last 5 years of three new adenomas, located in segments II and III and at the tip of the left liver ranging from 10 to 16 mm. These are presently uncomplicated, without any sign of necrosis, needing semestrial follow-up.

The metabolic and hemorrhagic complications in this patient, as well as the frequent use of preventive protocols in other platelet dysfunctions with elevated BT (Erstad 2001; Federici 2008; Kozek-Langenecker 2007) and the known improvement in BT after desmopressin (Marti et al. 1986), prompted us to adjust the perioperative protocol aiming at preventing metabolic decompensation and bleeding linked with specific hemostatic disorders in GSD patients.

The protocol is described below: BT and platelet function tests were evaluated at the time of admission for basis analysis, normal Ivy time (3 points puncture) being less than 8 min and normal platelet aggregation obtained on PFA-100 as occlusion times col-ADP < 120 s and col-adrenalin < 160 s.

A central jugular catheter was placed 5 days before surgery and glucose infusion was started at a rate of 4 mg/kg/min while following a daily cornstarch diet but without nightly tube feeding. Additional management included 1 mg/kg/day (maximum 60 mg/day) intravenous corticoid treatment (Solumedrol-Methylprednisolone), 3 days before surgery. Desmopressin (1 deamino-8-D-arginine

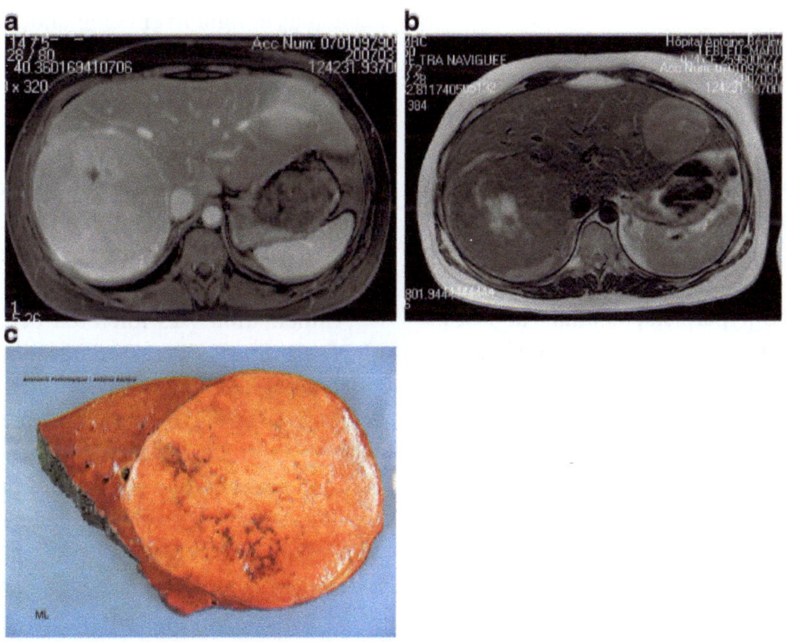

Fig. 1 Correlation of preoperative imaging and gross pathology of liver adenomatosis in patient 4. (**a**): Venous phase tomodensitometry of major adenoma of 11 cm located over segments VI-VII-VIII spontaneously hypodense with necrotic center and infracentimetric calcification at its pole. (**b**): Arterial uptake without late wash-out in the same adenoma. (**c**): Section of adenoma, the voluminous tumor has clear cut section without capsule

Table 3 Haemostasis results for all patients

Patient	1	2	3	4	5	6
Ivy BT (min) N<8 min	17	4	>30	No data	20	Not done
PFA col ADP (s) N 95–120 s	103	>300	>300	No data	141	>300/>286*
PFA col-adrenalin (s) N 95–160 s	133	>300	>300	No data	180	>248/193
Correction after glucose infusion	No data	249/247**	No data	No data	BT 5 min	187/161**

*Before/after interruption of salicylate
**PFA colADP/col-adrenalin

vasopressin DDAVP) was then injected (30 min slow infusion at 0.3 µg/kg), 1 h before surgical time. This injection could be repeated 12 and 24 h later if any unusual bleeding occurred postoperatively. Water restriction was prescribed during 24 h while increasing glucose supply to 6 mg/kg/min. One day after surgery, oral medication with tranexamic acid (Exacyl) followed the desmopressin prescription for five more days (1,500 mg/day or 20 mg/kg/day for children).

During surgery, glucose and lactic acid were sampled every hour to control metabolic balance. Postoperatively, glucose parenteral infusion was decreased by 1 mg/kg/min every 8 h if oral nutrition was tolerated while closely controlling the same biological parameters. We report below the results of five consecutive patients undergoing hepatic resection using this protocol with an adaptated protocol for the latest.

Cases 2 to 6

The age at diagnosis of GSDIa ranged from 5 to 60 months (mean 23.6 months), age at which frequent hyperglucidic meals were prescribed as well as nocturnal gastric drip feeding for all but patient 6 until puberty. All patients had annual follow-up including hepatic ultrasonography or more recently magnetic resonance imaging and biological markers of disease. For all patients but last compliance to treatment and diet were lacking before diagnosis of adenomas with frequent hypoglycemia and hyperlactacidemia. The last patient had chronic hyperlactatemia without detected hypoglycemia despite good familial compliance with the regimen. The mean age at first diagnosis of asymptomatic hepatic adenomas on ultrasonography was 14.5 years (range 13–16.25 years), the major adenoma at that time ranging from 0.8 to 3.6 cm (mean 2.44 cm). The age at surgery ranged from 18 to 30 years (mean 23 years). Main adenoma at the time of surgery ranged from 8 to 14 cm (mean 10.9 cm).

Indication for surgical treatment was linked with chronic or abdominal pain related to intratumoral complication such as central necrosis or hemorrhage for three patients and to rapid volume evolution in a few months leading to suspicion for malignant transformation for two.

Hemostatic disorder was confirmed before surgery for all patients (Table 3) except for patient number 4 who was transferred during a weekend from an outclinic cardiologic intensive care unit to surgery for abdominal pain mimicking pericarditis. She thus benefited from the complete accelerated protocol (glucose and corticoid treatment begun simultaneously) without anterior evaluation of PFA.

Patient number 6 benefited from a personalized protocol in regard to his other conditions.

One patient benefited from preoperative blood transfusion 1 month before partial hepatectomy for chronic anemia under 8 g/dl resistant to iron supplements. Another patient benefited from the whole protocol comprising desmopressin despite normal platelet aggregation times as BT was quite prolonged.

The patients had partial hepatectomy depending on the location of adenomas, five having right hepatectomy, associated for two patients with atypical resection of segments III and/or IV and for one patient with radiofrequency treatment under ultrasonography control on a distant and inaccessible nodule in segment II. One patient had a left hepatectomy, all adenomas in right liver being under 2 cm diameter at the time of surgery. No hemorrhage, metabolic, or intraoperative complication was deplored for the patients, bleeding being controlled under less than 1 l for all. One patient had a gas embolism that occurred during the dissection of the medial hepatic vein, followed by a profound decrease in PCO2 necessitating ephedrine and noradrenaline. She had acinetobacter postoperative peritonitis necessitating reoperation and intravenous antibiotics, thus prolonging hospital stay.

Alimentation was tolerated before day three for all and glucose infusion decreased consequently.

Histological analysis confirmed the diagnosis of benign adenomas for all. Further outcome is good with follow-up ranging from 8 to 35 months, with stable remaining adenomas for one, absence of residual or new adenomas for two, significant growth for one from 2 to 7.8 cm in 1 year, and one last patient who refuses all assessment since surgery but has no clinical complaints.

In patient 6, the use of the hemostatic protocol was adapted to the associated MoyaMoya disease (Goutières et al. 1997). Basic PFAs for this patient were increased to >300 and >248 s, possibly linked with chronic intake of salicylic acid to prevent cerebral infarction. The treatment was disrupted 3 weeks before surgery and hemostatic control showed improved but still increased PFA times (Table 3). Despite a new important anastomotic network after bilateral revascularization procedures by encephaloduroarterio-myo-synangiosis, desmopressin, and water restriction remained contraindicated for this patient with elevated risks of cortical softening. Thus, he received platelet transfusion at anesthetic induction, but bleeding occurred up to 900 ml necessitating transfusion of 1 RBC unit without macromolecular filling for maintaining normal blood pressure.

He had a prolonged hospital stay for nutrition purposes and for the first introduction of gastric drip fluid to maintain energy intake above 1,500 kCal/day.

Liver and GSD Patients

During the evolution of GSDIa, hepatic complications may appear such as focal nodular hyperplasia, hepatocellular adenoma, focal fatty infiltration, and hepatocellular carcinoma (Rake 2002a, b, Lee 2002, Labrune 1997, Labrune et al. 2002). The reported prevalence of adenomas ranges from 16 to 75%, and they mostly appear during or after puberty (Bianchi 1993; Chen et al. 1993; Lee 2002; Rake et al. 2002a, Talente et al. 1994). The mean age of first diagnosis of adenoma in our six patients is 14.46 years, and the progression to surgery is relatively long (9.71 years), in agreement with the ESGSDI which reported that prevalence of HCAs increases with age such that 70–80% of patients above 25 years have at least one lesion and that progression in size and/or number occurs in 50% of patients (Rake et al. 2002a). This does not differ from the rest of our GSDI population in whom 70.6% present HCAs with a mean age at first diagnosis of 20.9 years (range: 10–35). The patients without HCAs range from 11 to 30 years with a mean age of 19 years.

Stabilization or even regression of tumors has been described under strict metabolic homeostasis (Parker et al. 1981), but was not observed in our six patients, mostly poorly compliant as prescribed regimens were not observed in four out of six patients leading to chronic hypoglycemia and frequent lactatemia above 3 mmol/l (results in Table 1). In a recent study on GSD Ib mice, Yiu et al. demonstrated that normoglycemia and long-term metabolic correction alone do not prevent the apparition of hepatic tumors. Despite having maintained glucose homeostasis throughout the study, all mice accumulated excessive hepatic glycogen and fat; 40% developed steatohepatitis and multiple adenomas with one undergoing malignant transformation (Yiu et al. 2009). Thus, these nodules may appear in an even larger number of GSD patients through the years despite an improvement in daily management.

In our series, the youngest two male patients presented a very rapid evolution with doubling of tumor each year from the diagnosis of the first adenoma to surgery (less than 4 years). One patient was on a growth hormone treatment for severe growth retardation and partial growth hormone deficiency on biological data (data not shown) with pubertal delay. Treatment showed absence of efficacy on growth or pubertal start. The other patient used testosterone, creatine, and protein supplements in an unapproved bodybuilding program due to unease with his truncal obesity caused by glycogenosis. The oldest male patient also used male testosterone hormones in a parental attempt to induce growth, although the patient did achieve his expected genetic height despite final size at −2SD, thus increasing the risk of adenomas and liver adenomatosis in an otherwise at-risk population as adenomas can occur in men who take anabolic steroids (Klatskin 1977). Thus, we emphasize the contraindication for any hormonal treatment in GSD patients. In gynecologic management of GSDI female patients, guidelines notify a contraindication for ethinyloestradiol, because of the risks of hepatic adenoma and hypertriglyceridemia, and recommend progestative pills (Mairovitz et al. 2002). Female patients are compliant to these recommendations as all three women were on progestogen pills but the message was not conveyed to male patients or family as both used hormones as automedication.

Thus, adenomas must be closely followed as complications may occur such as abdominal pain, acute necrosis, intra-tumoral bleeding or intra-peritoneal rupture for superficial tumors and rare malignant transformation (Franco et al. 2005). Risk of hemorrhage (intra-tumoral bleeding) or rupture seems to be directly related to tumor size (Ribeiro et al. 1998; Veteläinen et al. 2008), becoming major above 5-cm diameter (Deneve et al. 2009). In our reported patients, one suffered from intra-tumoral hemorrhage without acute anemia or intra-peritoneal rupture, despite adenomas above 5 cm diameter. Painkillers such as morphinic drugs helped postpone surgical management. In the remainder cohort with adenoma and without surgical treatment, two were found to have intratumoral bleeding on systematic RMN without any previous abdominal pain reports.

Surgical Management

Surgical management of GSD patients with liver adenomatosis remains complex, due to the large number of adenomas, which require major liver resection on the one hand and the risks of hemorrhagic complications on the other. There is no systematic proposal for surgery in our medical center, neither for size nor for imaging criteria; patients with stable large adenomas above 10 cm diameter are actually carefully and regularly checked. Thus, indications for surgical removal of tumors in our six patients were classical (Chiche et al.

2000): acute complications such as hemorrhage or sudden necrosis causing asthenia or pain for four patients and volume evolution before the appearance of any complication for the last two (Table 2).

Partial hepatectomy or simple resection of adenoma might become a usual surgical procedure for GSD patients, especially if tumor size above 5 cm becomes a formal indication for surgical removal (Deneve et al. 2009). Currently, there are no guidelines on the optimal management of diffuse hepatocellular adenomatosis. Surgical removals of large or complicated nodules in symptomatic patients have been shown to significantly improve symptoms (Deneve et al. 2009; Chiche et al. 2000). The surgical dilemma that arises in these patients with multiple lesions is which mass to resect and which to leave behind, because all segments of the liver may be involved and partial resection may lead to an increased regeneration of remaining segments with the consequent progression of tumors left in situ. Our third patient shows a profile of rapid postoperative progression of nodules, without any hormonal or metabolic explanation for such a progression, whereas other patients have a more stable profile. Other conservative procedures with lesser risk of bleeding such as radiofrequency ablation (RFA) must be evaluated. This technique, which destroys liver tumors up to 3 cm in situ by localized application of heat to produce coagulative necrosis (Iannitti et al. 2002; Fujita et al. 2006; Rocourt et al. 2006), was successfully used for a posterior mass not accessible during partial hepatectomy in one of our patients. Adenoma was totally excluded at 1 and 6 months with a millimetric hypodense zone as the only sequel. Adjunction treatment to ablating all identifiable disease reduces the potential for malignant transformation or subsequent complications such as intraperitoneal hemorrhage (Fujita et al. 2006) in the long term, reducing still the risk of fulminating progression of remaining adenomas. For patient 5, right hepatectomy associated with atypical resection of left hepatic adenomas in segment IV was performed as well as radiofrequency treatment under ultrasonography control on a distant inaccessible nodule in segment II. To our knowledge, radiofrequency pulse ablation was used for the first time on an hepatic adenoma for a GSDI patient, while already discussed and used in tumors from other origins (oral contraceptives, exogenous androgens, hepatocarcinoma, metastatic nodules) (De Baere et al. 2000; Iannitti et al. 2002; Rocourt et al. 2006). It was also successfully used on the contralateral side from hepatic resection to permit the management of all detectable disease in patients with multiple adenoma of diffuse distribution presentation reminiscent of that of GSD patients (Fujita et al. 2006). No recurrence was found at ablation sites over 10 years (Deneve et al. 2009) and no trace of adenoma is visible in our patient.

Partial hepatectomy for patients with widespread adenomas as practiced in our institute seems to be an alternative to hepatic transplantation advocated today for GSD patients, as adenoma resection is promoted as awaiting therapy before transplantation. Liver transplantation once appeared ideal for multiple, diffuse, and voluminous adenomas in GSD as it involves a complete tumoral resection with no ulterior risk of expansion and corrects the underlying liver dysfunction (Ji et al. 2009; Labrune 2002; Lerut et al. 2003). Duration of graft, shortage of donors, necessary life long treatments against rejection become mandatory in these poorly compliant patients thus creating new problems and the chronic use of steroid treatments and other immunosuppressive drugs need to be balanced with the possibility of partial hepatectomy in such patients.

Bleeding and Metabolic Managements

Although likely to appear during all hepatic surgery, these bleeding risks are considerably increased in GSD population by a defect in platelet aggregation, which can occur in up to 23% of GSDI patients (Rake et al. 2002a). PFA explorations in our other 12 patients of our GSD population show increased occlusion times in 92% (11/12) of patients whether or not metabolic control is good (data not shown). Hemostatic explorations are mandatory in this pathology before any surgical procedure, but it appears, in our patients, that bleeding tendency may persist although in vitro platelet function aggregation status is in the normal range (patient 1).

In this rare disease, there are only few reports on perioperative management for partial hepatectomy or other scheduled surgery (Rake et al. 2002b; Oshita et al. 2008; Reddy et al. 2007). The literature about normalized hemostatic or metabolic parameters suggest a complete benefit of continuous glucose infusion after 1 week in one patient (Hutton et al. 1976) and the recurrence of bleeding diathesis after discontinuation of treatment. Glucose infusion was justified by the link between bleeding diathesis and chronic hypoglycemia. In our patients, bleeding was only partially contained by 5 days of continuous steady glucose infusion We propose a preventive protocol concerning hemostasis using systematic corticoid and desmopressin. Although transient possible accumulation of glycogen in hepatocytes might appear after short-term high doses of steroids (Iancu et al. 1986), we used corticoid treatment in addition to usual glucose infusion before surgery as elevated liver synthesis of hemostatic factors exists in GSD patients (Labrune et al. 1994) and persists after glucose infusion (Marfaing-Koka et al. 2003). Corticoids permit specific inhibition of the synthesis of prostaglandins and leucotriens via inhibition of Phospholipase A2 (Jobin

1995). They prevent bleeding if used locally in some otorhinologic surgeries (Albu et al. 2010). They also have neurological (Limbourg et al. 2002), cardiovascular (Hafezi-Moghadam et al. 2002) or renal protection (Baker et al. 2006) during surgery with potential ischemia. Enlargement of the liver was not observed in our population on gross clinical evaluation. Resected liver was steatotic but never described as having glycogen surcharge. Desmopressin treatment has already been found as effective in correcting BT as well as the von Willebrand antigen and activity in GSD patients (Marti et al. 1986), or used to control massive hemorrhage through hepatic surgery in the same population (Reddy et al. 2007). Frequent morbidity and mortality related to hemorrhage during hepatic resection are reported in GSD patients (Reddy et al. 2007) and this excessive morbidity needs prevention. DDAVP effect on hemostasis and safety in clinical use has been demonstrated in von Willebrand disease (Federici 2008), on other intrinsic platelet disorders, or after use of salicylate, and it is used during known hemorrhagic surgery in patients without bleeding disorders (Jobin 1995). DDAVP induces a release of VWF threefolds normal values in normal subjects and a secondary increase of factor VIII and has a direct effect on platelet membrane glycoproteins. This action is used in known hemorrhagic surgery such as vertebral arthrodesis in patients with normal hemostasis (Jobin 1995; Hashemi et al. 1993). In vitro experiments show an inhibition in the binding of radioactive factor VIII/von Willebrand's factor to platelet by various fractions of GSD-I plasma. DDAVP has been used during excessive bleeding and complicated surgery after bleeding in GSD patients (Reddy et al. 2007) but never preventively. Von Willebrand concentration is normal in GSD patients (Marfaing-Koka et al. 2003) and response recorded is identical to other patients. DDAVP also releases the tissue plasminogen activator and induces a transient fibrinolytic activity (Hashemi et al. 1993) which tranexamic acid counters. Its efficacy in high risk for hemorrhage situations has been demonstrated (Kozek-Langenecker 2007; Shakur et al. 2010).

The management of our patients who underwent partial hepatectomy covered by hemostatic protocol remained easy without abnormal bleeding and without increased morbidity related to hemorrhage.

While first patient had severe hemorrhage needing blood transfusion, macromolecular filling and ephedrine drugs, no other patient had any hemodynamic instability, nor any need for transfusion, vascular filling, or vasoconstrictor drugs after the introduction of this protocol.

For all four patients, the use of desmopressin was limited to first dose 30 min before surgery as, in absence of any hemorrhage, doses at H12 and H24 were not injected. Thus we propose this extended protocol for important interventions whether or not BT and PFA are in the normal range in GSD patients, and especially for poorly compliant patients with secondary bleeding diathesis, to limit morbidity linked to hemorrhage. In the case of patients with good compliance to treatment and good metabolic control, the benefits of assured limited hemorrhagic risk should be considered in time of shortage of blood units and putative risks of transfusion.

After dismissal from the hospital, one of the female patients underwent a surgical ovarian tumorectomy by coelioscopy for bilateral teratomas without the use of hemostatic protocol and suffered from large, painful, local hematomas and acute anemia prolonging hospital stay of more than 12 days. In our experience, even minor surgical approaches should benefit from previous hemostatic preparation with desmopressin, as some of our patients have suffered from large and painful hematomas for such procedures as teeth removal or abscess incision.

In summary, we present the perioperative care of six GSD type Ia patients undergoing partial hepatectomy for complicated or voluminous tumor masses. Of primary importance for the surgeons are issues related to the risk of continuous hemorrhage due to platelet dysfunction, and alterations in the metabolic regulation of glucose. All these concerns could be prevented via a perioperative protocol using 4–6 mg/kg/min glucose infusion, corticoid injections around surgical time associated with desmopressin and antifibrinolytic medication. RFA of a posterior mass was used with success combined with a contralateral hepatic resection, but more time is necessary to evaluate the risk of regeneration after such a technique in this rare disease.

References

Albu S, Gocea A, Mitre I (2010) Preoperative treatment with topical corticoids and bleeding during primary endoscopic sinus surgery. Otolaryngol Head Neck Surg 143:573–578

Ambruso DR, MacCabe ER, Anderson D, Beaudet A, Ballas LM, Brandt IK, Brown B, Coleman R, Dunger DB, Falletta JM et al (1985) Infectiuos and bleeding complications in patients with glycogenosis Ib. Am J Dis Child 139:691–697

Baker RC, Armstrong MA, Young IS, McClean E, O'Rourke D, Campbell FC, D'Sa AA, McBride WT (2006) Methylprednisolone increases urinary nitrate concentrations and reduces subclinical renal injury during infrarenal aortic ischemia reperfusion. Ann Surg 244:821–826

Bianchi L (1993) Glycogen storage disease 1 and hepatocellular tumours. Eur J Pediatr 152:S63–S70

Chen YT, Bazarre CH, Lee MM, Sidbury JB, Coleman RA (1993) Type 1 glycogen storage disease: nine years of management of cornstarch. Eur J Pediatr 152:S56–S59

Chiche L, Dao T, Salame E, Galais MP, Bouvard P, Schmutz G, Rousselot P, Bioulac-Sage P, Ségol P, Gignoux M (2000) Liver adenomatosis: reappraisal, diagnosis, and surgical management: eight new cases and review of the literature. Ann Surg 231:74–81

De Baere T, Elias D, Dromain C, Din MG, Kuoch V, Ducreux M, Boige V, Lassau N, Marteau V, Lasser P, Roche A (2000) Radio-

frequency ablation of 100 hepatic metastases with mean follow-up of more than 1 year. AJR Am J Roentgenol 175:1619–1625
Deneve JL, Pawlik TM, Cunningham S, Clary B, Reddy S, Scoggins CR, Martin RC, D'Angelica M, Staley CA, Choti MA, Jarnagin WR, Schulick RD, Kooby DA (2009) Liver cell adenoma: a multicenter analysis of risk factors for rupture and malignancy. Ann Surg Oncol 16:640–648
Erstad BL (2001) Systematic hemostatic medications for reducing surgical blood loss. Ann Pharmacother 35:925–934
Federici AB (2008) The use of desmopressin in von Willebrand disease: the experience of the first 30 years (1977–2007). Haemophilia 14(Suppl 1):5–14
Franco LM, Krishnamurthy V, Bali D, Weinstein DA, Arn P, Clary B, Boney A, Sullivan J, Frush DP, Chen YT, Kishnani PS (2005) Hepatocellular carcinoma in glycogen storage disease type 1a: a case series. J Inherit Metab Dis 28:153–162
Fujita S, Kushihata F, Herrmann GE, Mergo PJ, Liu C, Nelson D, Fujikawa T, Hemming AW (2006) Combined hepatic resection and radiofrequency ablation for multiple hepatic adenomas. J Gastroenterol Hep 21:1351–1354
Gilchrist GS, Fine RN, Donnell GN (1968) The hemostatic defect in glycogen storage disease, type 1. Acta Paediatr Scand 57:205–208
Goutières F, Bourgeois M, Trioche P, Demelier JF, Odievre M, Labrune P (1997) Moyamoya disease in a child with glycogen storage disease type Ia. Neuropediatrics 28:133–134
Hafezi-Moghadam A, Simoncini T, Yang Z, Limbourg FP, Plumier JC, Rebsamen MC, Hsieh CM, Chui DS, Thomas KL, Prorock AJ, Laubach VE, Moskowitz MA, French BA, Ley K, Liao JK (2002) Acute cardiovascular protective effects of corticosteroids are mediated by non-transcriptional activation of endothelial nitric oxide synthase. Nat Med 8:473–479
Hashemi S, Palmer DS, Ave MT, Ganz PR (1993) Platelet-activating factor secreted by DDAVP-treated monocytes mediates von Willebrand factor release from endothelial cells. Cell Physiol 154:496–505
Hutton RA, Macnab AJ, Rivers RP (1976) Defect of platelet function associated with chronic hypoglycaemia. Arch Dis Child 51:49–55
Iancu TC, Shiloh H, Dembo L (1986) Hepatomegaly following short-term high-dose steroid therapy. J Pediatr Gastroenterol Nutr 51:41–46
Iannitti DA, Dupuy DE, Mayo-Smith WW, Murphy B (2002) Hepatic radiofrequency ablation. Arch Surg 137:422–426
Ji HF, Wang WL, Shen Y, Zhang M, Liang TB, Wu J, Xu X, Yan S, Zheng SS (2009) Reduced-size liver transplantation for glycogen storage disease. Hepatobiliary Pancreat Dis Int 8:106–108
Jobin F (1995) L'hémostase. Presses Université Laval, Editions Maloine, Québec, Paris, p 496
Kao KJ, Coleman RA, Pizzo SV (1980) The bleeding diasthesis in human glycogen storage disease type I: in vitro identification of a naturally occurring inhibitor of ristocetin-induced platelet aggregation. Thromb Res 18:683–692
Klatskin G (1977) Hepatic tumors: possible relationship to use of oral contraceptives. Gastroenterology 73:386–394
Kozek-Langenecker S (2007) Management of massive operative blood loss. Minerva Anestesiol 73:401–415
Labrune P (2002) Glycogen storage disease type 1: indications for liver and/or kidney transplantation. Eur J Pediatr 161:S53–S55
Labrune P, Benattar C, Ammoury N, Chalas J, Lindenbaum A, Odièvre M (1994) Serum concentrations of albumin, C-reactive protein, alpha 2-macroglobulin, prealbumin, fibronectin, fibrinogen, transferrin, and retinol binding protein in 55 patients with hepatic glycogen storage diseases. J Pediatr Gastroenterol Nutr 18:41–44
Labrune P, Trioche P, Duvaltier I, Chevalier P, Odievre M (1997) Hepatocellular adenomas in glycogen storage disease type I and III: a series of 43 patients and review of the literature. J Pediatr Gastroenterol Nutr 24:276–279
Lee PJ (2002) Glycogen storage disease type 1: pathophysiology of liver adenomas. Eur J Pediatr 161:S46–S49
Lerut JP, Ciccarelli O, Sempoux C, Danse E, de Flandre J, Horsmans Y, Sokal E, Otte JB (2003) Glycogen storage type 1 diseases and evolutive adenomatosis: an indication for liver transplantation. Transpl Int 16:879–884
Limbourg FP, Huang Z, Plumier JC, Simoncini T, Fujioka M, Tuckermann J, Schütz G, Moskowitz MA, Liao JK (2002) Rapid nontranscriptional activation of endothelial nitric oxide synthase mediates increased cerebral blood flow and stroke protection by corticosteroids. J Clin Invest 110:1729–1738
Mairovitz V, Labrune P, Fernandez H, Audibert F, Frydman R (2002) Contraception and pregnancy in women affected by glycogen storage disease. Eur J Pediatr 161:S97–S101
Marfaing-Koka A, Boyer-Neumann C, Meyer D, Odievre M, Labrune P (2003) Increased levels of haemostatic proteins are independent of inflammation in glycogen storage disease type 1a. J Pediatr Gastroenterol Nutr 37:566–570
Marti GE, Rick ME, Sidbury J, Gralnick HR (1986) DDAVP infusion in five patients with type 1 glycogen storage disease and associated correction of prolonged bleeding times. Blood 68:180–184
Oshita A, Itamoto T, Amano H, Ohdan H, Tashiro H, Asahara T (2008) Perioperative management of benign hepatic tumors in patients with glycogen storage disease type Ia. J Hepatobiliary Pancreat Surg 15:200–203
Parker P, Burr I, Slonim A, Ghishan FK, Greene H (1981) Regression of hepatic adenomas in type 1a glycogen storage disease with dietary therapy. Gastroenterology 81:534–536
Rake JP, Visser G, Labrune P, Leonard JV, Ullrich K, Smit GPA (2002a) Glycogen storage disease type 1: diagnosis, management, clinical course, and outcome. Results of the European study on glycogen storage disease type 1 (ESGSDI). Eur J Pediatr 161: S20–S34
Rake JP, Visser G, Labrune P, Leonard JV, Ullrich K, Smit GPA (2002b) Guidelines for management of glycogen storage disease type 1 – European study on glycogen storage disease type 1 (ESGSDI). Eur J Pediatr 161:S112–S119
Reddy SK, Kishnani PS, Sullivan JA, Koeberl DD, Desai DM, Skinner MA, Rice HE, Clary BM (2007) Resection of hepatocellular adenoma in patients with glycogen storage disease type Ia. J Hepatol 47:658–663
Ribeiro A, Burgart LJ, Nagorney DM, Gores GJ (1998) Management of liver adenomatosis: results with a conservative surgical approach. Liver Transpl Surg 4:388–398
Rocourt DV, Shiels WE, Hammond S, Besner GE (2006) Contemporary management of benign hepatic adenoma using percutaneous radiofrequency ablation. J Pediatr Surg 41:1149–1152
Sahud MA (1972) Platelet disorders – a review of disturbances in adhesion, aggregation, and release reaction. Calif Med 116: 21–31
Shakur H, Elbourne D, Gülmezoglu M, Alfirevic Z, Ronsmans C, Allen E, Roberts I (2010) The WOMAN trial (World Maternal Antifibrinolytic Trial): tranexamic acid for the treatment of postpartum haemorrhage: an international randomised, double blind placebo controlled trial. Trials 11:40
Talente GM, Coleman RA, Alter C, Baker L, Brown BI, Cannon RA et al (1994) Glycogen storage disease in adults. Ann Intern Med 120:218–226
Veteläinen R, Erdogan D, de Graaf W, ten Kate F, Jansen PL, Gouma DJ, van Gulik TM (2008) Liver adenomatosis: re-evaluation of aetiology and management. Liver Int 28:499–508
Yiu WH, Pan CJ, Mead PA, Starost MF, Mansfield BC, Chou JY (2009) Normoglycemia alone is insufficient to prevent long term complications of hepatocellular adenoma in glycogen storage disease type 1b mice. J Hepatol 51:909–917. doi:10.1016/j.jhep. 2008.11.026

JIMD Reports
DOI: 10.1007/8904_2011_24

Received: 29 December 2010; Revised: 1 February 2011;
Accepted: 3 February 2011; Published online: 22 June
Communicated by: Georg Hoffmann

Newborn Screening for Tyrosinemia Type I: Further Evidence that Succinylacetone Determination on Blood Spot Is Essential

Giancarlo la Marca, Sabrina Malvagia, Elisabetta Pasquini, Catia Cavicchi, Amelia Morrone, Federica Ciani, Silvia Funghini, Fabio Villanelli, Enrico Zammarchi, and Renzo Guerrini

Abstract Tyrosinemia type I is a genetic disorder characterized by accumulation in the blood and urine of the toxic metabolite succinylacetone (SUAC), not detectable in healthy samples. In many countries, newborns are screened for tyrosinemia type I using tyrosine as a primary marker. Unfortunately, tyrosine accumulation may take longer to occur and it may be not obvious when specimens are collected, in the first few days of life, as for newborn screening. In 2008, we reported changes to simultaneously measure acylcarnitines, amino acids, and SUAC during expanded newborn screening. We established the usefulness of this method after identifying a first asymptomatic newborn affected by tyrosinemia type I. Now we report a second infant with positive SUAC screening result (14.1 μmol/L, n.v.<2) and normal tyrosine concentration (74 μmol/L; n.v.<250). We also performed molecular analysis of *FAH* gene in both patients after diagnosis at newborn screening. They had consanguineous parents and were both homozygous for two known disease-causing mutations of the *FAH* gene. The outcome of patients detected in the MS/MS screening is significantly favorable. We also report our results of newborn screening for tyrosinemia type I before and after inclusion of SUAC as a primary marker for this disease.

Keywords Newborn screening · Succynilacetone · Tyrosinemia type I

Introduction

Tyrosinemia type I (MIM 276700) affects 1 in every 100,000 to 120,000 babies worldwide, although the real incidence could be higher. It is an autosomal recessive disorder caused by mutations in the *FAH* gene that leads to deficiency of the fumarylacetoacetic hydrolase (FAH; EC 3.7.1.2), the last enzyme in the tyrosine degradation pathway. The consequence of the metabolic block is the conversion of the catabolic

G. la Marca (✉)
Mass Spectrometry Laboratory, Clinic of Pediatric Neurology, A. Meyer Children's Hospital, Florence, Italy
and
Department of Pharmacology, University of Florence, Florence, Italy
and
Neurometabolic Unit, Department of Pediatric Neurosciences, "A. Meyer" University Children's Hospital, Viale Pieraccini, 24, 50134 Florence, Italy
e-mail: g.lamarca@meyer.it, giancarlo.lamarca@unifi.it

S. Malvagia and S. Funghini
Mass Spectrometry Laboratory, Clinic of Pediatric Neurology, A. Meyer Children's Hospital, Florence, Italy
and
Neurometabolic Unit, Department of Pediatric Neurosciences, "A. Meyer" University Children's Hospital, Viale Pieraccini, 24, 50134 Florence, Italy

E. Pasquini, C. Cavicchi, and F. Ciani
Neurometabolic Unit, Department of Pediatric Neurosciences, "A. Meyer" University Children's Hospital, Viale Pieraccini, 24, 50134 Florence, Italy

A. Morrone
Neurometabolic Unit, Department of Pediatric Neurosciences, "A. Meyer" University Children's Hospital, Viale Pieraccini, 24, 50134 Florence, Italy
and
Department of Pediatrics, University of Florence, Florence, Italy

F. Villanelli
Mass Spectrometry Laboratory, Clinic of Pediatric Neurology, A. Meyer Children's Hospital, Florence, Italy

E. Zammarchi
Department of Pediatrics, University of Florence, Florence, Italy

R. Guerrini
Mass Spectrometry Laboratory, Clinic of Pediatric Neurology, A. Meyer Children's Hospital, Florence, Italy
and
Neurometabolic Unit, Department of Pediatric Neurosciences, "A. Meyer" University Children's Hospital, Viale Pieraccini, 24, 50134 Florence, Italy
and
Department of Neurology and Neurosurgery, "A. Meyer" University Children's Hospital, Florence, Italy

intermediates maleylacetoacetate and fumarylacetoacetate to the toxic metabolites SUAC and succinylacetoacetate. The presence of SUAC in urine or blood is pathognomonic for tyrosinemia type I (Mitchell et al. 2001), which, if untreated, is usually fatal within age 10 years. Affected children may exhibit diarrhea, vomiting, jaundice, liver or kidney failure, neurological crisis, rickets, failure to thrive, and hepatocellular carcinoma (Mitchell et al. 2001). With treatment, many patients can lead normal lives with few restrictions. Treatment usually involves 2-(2-nitro-4-3 trifluoro-methylbenzoyl)-1,3-cyclohexanedione (NTBC) administration and a diet low in tyrosine and phenylalanine (Lindstedt et al. 1992). Early diagnosis and initiation of therapy may thus be crucial in improving outcome (Joshi and Venugopalan 2004).

In many countries, newborns are screened for tyrosinemia type I using tyrosine as a primary marker (US Department of Health and Human Services, Maternal and Child Health Bureau 2005). An elevated concentration of tyrosine, however, is not sensitive enough to detect all cases (Wilcken et al. 2003). Some newborn screening programs have recently introduced the determination of SUAC as a reliable and valid marker for the identification of tyrosinemia type I (Allard et al. 2004; Sander et al. 2006; la Marca et al. 2008, Turgeon et al. 2008; Al-Dirbashi et al. 2008, Adam et al. 2009).

In 2008, we reported an efficient method to simultaneously measure acylcarnitines, amino acids, and SUAC during expanded newborn screening (la Marca et al. 2008). We established the usefulness of this method after identifying a first asymptomatic newborn with by tyrosinemia type I (la Marca et al. 2009). We now present a second infant with positive SUAC screening and tyrosine within the normal range. We also report our results of newborn screening for tyrosinemia type I before and after inclusion of SUAC as a primary marker for this disease.

Case Report

The patient was a Moroccan boy, born at full term after an uneventful pregnancy and delivery with a weight of 3,620 g. He was the first child of first degree cousins. Both parents were healthy, but the family reported a positive history for deaths (three babies) in early childhood of unknown causes. Blood for newborn screening, collected on the third day of life, revealed an elevated SUAC level of 14.1 μmol/L (n.v. <2). Tyrosine value was normal (74 μmol/L; n.v. <250). On the sixth day of life, the boy was hospitalized for further testing. The diagnosis of tyrosinemia type I was confirmed by detection of SUAC in the urine (70 mmol/mol of creatinine) and plasma (11 μmol/L). Plasma tyrosine level was increased (791 μmol/L; n.v. <123). The child was in good condition and no clinical manifestations were apparent. Liver transaminases, alkaline phosphatase, bilirubin, coagulation factors, and ammonia were within normal levels. NTBC treatment was started at a dose of 1 mg/kg/die body weight combined with dietary restriction of tyrosine and phenylalanine.

Results and Discussion

In many countries (Canada, USA, Latin America), to reduce hospital costs, newborn screening specimen collection is done within 24–48 h after birth in response to the early discharge of mothers and their infants. This tendency causes a serious problem for metabolic newborn screening since accumulation of some metabolites, used as markers, may occur only in some days after delivery. The success of newborn screening depends on the type of marker used. Hence, the challenge is to find and to include in newborn screening programs appropriate diagnostic markers for the early detection of metabolic disorders.

Up to now, many newborn screening programs worldwide use tyrosine levels as a marker for tyrosinemia type I. Unfortunately this metabolic condition cannot be detected when specimens are collected in the first few days of life, as tyrosine accumulation may take longer to occur (Mitchell et al. 2001). The increased SUAC levels, the metabolite immediately upstream of the enzymatic block, is a reliable, appropriate, and early marker of this disorder.

The case presented here, as also the previously reported case (la Marca et al. 2009), provides further evidence on the importance of using SUAC as the primary metabolic marker for detecting tyrosinemia type I in newborn screening.

Experience in various laboratories, including our own, suggests that tyrosine was not a sensitive diagnostic marker for the timely identification of patients and decreased the specificity of the test while increasing the unnecessary medical cost for false-positive recall rate. Indeed, the most common cause of increased blood tyrosine levels is benign transient tyrosinemia of the newborn.

The inclusion of SUAC in our newborn screening program dates back to January 2007, subsequently to a false-negative result. Among about 136,000 newborns screened, two patients with tyrosinemia type I were identified. No false positive was on record. Both patients had consanguineous parents (from Turkey and Morocco) and were homozygous for the known c.709C>T (p.Arg237X) and IVS6-1G>T disease-causing mutations of the *FAH* gene (Rootwelt et al. 1996; Ploos van Amstel et al. 1996). The IVS6-1G>T leads to a splicing defect and has been reported as common in the European and Mediterranean area (Arranz et al. 2002). The results for tyrosinemia type I in the Tuscan newborn screening program are reported in the Tables 1 and 2.

Table 1 Expanded newborn screening for tyrosinemia type I in Tuscany before and after inclusion of SUAC as a primary marker

	Newborn examined	Cases diagnosed	Cases false negatives
Before SUAC inclusion	113,090	0	1
After SUAC inclusion	136,075	2	0
Total	249,165	2	1

Table 2 Patients with tyrosinemia type I born in Tuscany between 2006 and 2010

	Ethnicity	Tyrosine, μmol/L (n.v. < 250)	SUAC, μmol/L
Patient 1[a]	Moroccan	142	14.9[b]
Patient 2	Turkish	126	7.6
Patient 3	Moroccan	74	14.1

[a]False-negative result on newborn screening before SUAC inclusion
[b]Original newborn screening card was analyzed 2 years after collection

It is very difficult to estimate the incidence of this disease in the population but, based on our own experience, we are inclined to believe that it is generally underestimated. Furthermore, in Turkey as well as Morocco and other North African and Arabic countries, the elevated rates of consanguinity may have an impact on the incidence of rare autosomal recessive disorders (Al-Gazali et al. 2006; Jaouad et al. 2009).

There are various methods for SUAC determination on DBS, with significant modifications in the newborn screening procedures. The method we had previously proposed (la Marca et al. 2008) consists of a simple and fast protocol for newborn screening sample preparation and allows identifying this metabolic defect in the neonatal period with 100% sensitivity. No cost for additional equipment or staff members is required for applying such testing.

Synopsis

Succinylacetone determination on dried blood spot.

References

Adam BW, Lim TH, Hall EM, Hannon WH (2009) Preliminary proficiency testing results for succinylacetone in dried blood spots for newborn screening for tyrosinemia type I. Clin Chem 55:2207–2213

Al-Dirbashi OY, Rashed MS, Jacob M, Al-Ahaideb LY, Al-Amoudi M, Rahbeeni Z, Al-Sayed MM et al (2008) Improved method to determine succinylacetone in dried blood spots for diagnosis of tyrosinemia type I using UPLC-MS/MS. Biomed Chromatogr 22:1181–1185

Al-Gazali L, Hamamy H, Al-Arrayad S (2006) Genetic disorders in the Arab world. BMJ 333:831–834

Allard P, Grenier A, Korson MS, Zytkovicz TH (2004) Newborn screening for hepatorenal tyrosinemia by tandem mass spectrometry: analysis of succinylacetone extracted from dried blood spots. Clin Biochem 37:1010–1015

Arranz JA, Piñol F, Kozak L, Pérez-Cerdá C, Cormand B, Ugarte M, Riudor E (2002) Splicing mutations, mainly IVS6–1(G>T), account for 70% of fumarylacetoacetate hydrolase (FAH) gene alterations, including 7 novel mutations, in a survey of 29 tyrosinemia type I patients. Hum Mutat 20:180–188

Jaouad IC, Elalaoui SC, Sbiti A, Elkerh F, Belmahi L, Sefiani A (2009) Consanguineous marriages in Morocco and the consequence for the incidence of autosomal recessive disorders. J Biosoc Sci 41:575–581

Joshi SN, Venugopalan P (2004) Experience with NTBC therapy in hereditary tyrosinemia type I: an alternative to liver transplantation. Ann Trop Paediatr 24:259–265

la Marca G, Malvagia S, Pasquini E, Innocenti M, Fernandez MR, Donati MA, Zammarchi E (2008) The inclusion of succinylacetone as marker for tyrosinemia type I in expanded newborn screening programs. Rapid Commun Mass Spectrom 22:812–818

la Marca G, Malvagia S, Funghini S, Pasquini E, Moneti G, Guerrini R, Zammarchi E (2009) The successful inclusion of succinylacetone as a marker of tyrosinemia type I in Tuscany newborn screening program. Rapid Commun Mass Spectrom 23:3891–3893

Lindstedt S, Holme E, Lock EA, Hjalmarson O, Strandvik B (1992) Treatment of hereditary tyrosinemia type I by inhibition of 4-hydroxyphenylpyruvate dioxygenase. Lancet 340:813–817

Mitchell GA, Grope M, Lambert M, Tanguay RM (2001) Hypertyrosinemia. In: Scriver CR, Beaudet AL, Sly WS, Valle D (eds) The metabolic and molecular bases of inherited disease. McGraw-Hill, New York, pp 1777–1805

Ploos van Amstel JK, Bergman AJ, van Beurden EA, Roijers JF, Peelen T, van den Berg IE, Poll-The BT, Kvittingen EA, Berger R (1996) Hereditary tyrosinemia type 1: novel missense, nonsense and splice consensus mutations in the human fumarylacetoacetate hydrolase gene; variability of the genotype-phenotype relationship.Hum Genet 97:51–59

Rootwelt H, Høie K, Berger R, Kvittingen EA (1996) Fumarylacetoacetase mutations in tyrosinaemia type I. Hum Mutat 7:239–243

Sander J, Janzen N, Peter M, Sander S, Steuerwald U, Holtkamp U, Schwahn B et al (2006) Newborn screening for hepatorenal tyrosinemia: tandem mass spectrometric quantification of succinylacetone. Clin Chem 52:482–487

Turgeon C, Magera MJ, Allard P, Tortorelli S, Gavrilov D, Oglesbee D, Raymond K et al (2008) Combined newborn screening for succinylacetone, amino acids, and acylcarnitines in dried blood spots. Clin Chem 54:657–664

US Department of Health and Human Services, Maternal and Child Health Bureau (2005) Newborn screening: toward a uniform screening panel and system report for public comment. Fed Regist 70:308–329

Wilcken B, Wiley V, Hammond J, Carpenter K (2003) Screening newborns for inborn errors of metabolism by tandem mass spectrometry. N Engl J Med 348:2304–2312

JIMD Reports
DOI: 10.1007/8904_2011_25

Received: 5 January 2010; Revised: 4 January 2011;
Accepted: 13 January 2011; Published online: 25 June
Communicated by: Georg Hoffmann

Utility of Rare Disease Registries in Latin America

Ana Maria Martins, Marcelo Kerstenezky, Adriana Linares, Juan Politei, Regina Kohan, Sandra Ospina, Carmen Varas, Jacobo Villalobos, Hernán Amartino, Sergio Franco, Guilhermo Valadez, Roberto Giugliani, Patricio Guerra, and Luz Sanches

A.M. Martins (✉)
Fabry Registry Brazil, Latin America Fabry, Gaucher, MPS I and Pompe Registries, Universidade Federal de São Paulo, São Paulo, Brazil
e-mail: ana.martins@unifesp.br

M. Kerstenezky
Gaucher Registry Brazil, Hospital da Restauração, Recife, Pernambuco, Brazil

A. Linares
Gaucher Registry Colombia, Universidad Nacional de Colombia, Bogotá, Colombia

J. Politei
Fabry Registry Argentina, Hospital Juan Fernández, Buenos Aires, Argentina

R. Kohan
Registry Gaucher Argentina, Hospital Ramos Mejía, Buenos Aires, Argentina

S. Ospina
Fabry, MPS I and Pompe Registries Colombia, Universidad del Rosario, Bogotá, Colombia

C. Varas
Fabry Registry ChileHospital San Pablo, Coquimbo, Chile

J. Villalobos
Gaucher, Fabry, MPS I and Pompe Registries Venezuela, Universidad Central de Venezuela, en Caracas, Venezuela

H. Amartino
Pompe Registry Argentina, Hospital Universitario Austral, Buenos Aires, Argentina

S. Franco
Gaucher Registry Mexico, Instituto Mexicano del Seguro Social, Mexico City, Mexico

G. Valadez
Fabry Registry Mexico, Instituto Mexicano del Seguro Social, Mexico City, Mexico

R. Giugliani
MPS I Registry Brasil, Universidade Federal do Rio Grande do Sul, Porto Alegre, Brazil

P. Guerra
MPS I and Pompe Registries Chile, Hospital de Puerto Montt-Universidad San Sebastián, Concepción, Chile

L. Sanches
MPS I Registry, Hospital de Especialidades, Instituto Mexicano del Seguro Social, Mexico City, Mexico

Abstract There are many registries in Latin America as dialysis and kidney transplantation, breast cancer, primary immunodeficiency, acute coronary syndromes, but the focus here are the registries of lysosomal storage diseases (LSD) because is our experience. Registry of Gaucher disease, Fabry disease, Pompe disease, and mucopolysaccharidosis type I are comprehensive observational voluntary programs that aim to collect clinical and laboratory data of initiation, progression, and evolution of those diseases, with and without treatment, using questionnaires of quality of life and/or skills and functions. There are two more programs of LSD: Hunter outcome survey and Fabry outcome survey. The registries are a kind of phase IV clinical trials, postmarketing studies delineate additional information including the drug's risks, benefits, and optimal use, and in addition we have data from natural history. The demographics of the Gaucher, Fabry, MPS I, and Pompe Registries show that a total of patients, being 16%, 8%, 15%, and 7%, respectively, of this population, and 19%, 19%, 18%, and 13%, respectively, of all physicians participating in the program are from Latin America. In the Gaucher Registry, we can observe that the percentage of children in Latin America (29%) is bigger than the rest of the world (20%), what can mean more severe disease in this population. These diseases are rare, and a database of clinical data from a larger number of patients gives us the opportunity to know about the natural history of these diseases, their phenotypic variability, and the response to specific enzyme replacement therapy in our population.

Keywords Disease registries · Fabry registry · Gaucher registry · Latin America diseases registries · Mucoolisaccharidosis I registry · Pompe registry

Introduction

Lysosomal storage diseases (LSD) including in Latin America, we have registries of Gaucher disease, Fabry disease, Pompe disease, mucopolysaccharidosis type I, Hunter outcome survey

(HOS), and Fabry outcome survey (FOS). The infrastructure of the Gaucher Registry, Fabry Registry, Pompe Registry, and mucopolysaccharidosis type I (MPS I Registry) is sponsored by Genzyme Corporation (Cambridge, MA, USA), and they gave us permission to publish the registries data from 2009. All these registries are comprehensive observational voluntary programs that aim to collect clinical and laboratory data of initiation, progression, and evolution of those diseases, with and without treatment, using questionnaires of quality of life and/or skills and functions. The privacy of doctors and patients is protected, and they are identified only by the initials of their names.

The Advisory Boards are formed by a body of independent physicians specialized in each disease, and they coordinate the regular meetings of those programs to assess the possibilities for scientific research; evaluate the quality and specificity of the collected data, making changes in these data whenever it is necessary; analyze the request of the registries data for publications by the participants; schedule the publications; create working groups among the participating physicians; and evaluate consultations on the registries.

The above-mentioned diseases are rare, and a database of clinical data from a larger number of patients gives us the opportunity to know more about the natural history of these diseases, their phenotypic variability, and the response to enzyme replacement therapy. Often patients present clinical manifestations that were not previously described in the literature. The physician participating in the program can query the registry database of that particular disease, questioning whether the observed event is recorded and, according to the response, the doctor may use this information for a publication or presentation at medical meetings, reverting the findings into a better knowledge about the disease.

The coordinators of the registries of LSD in Latin America have an educational role with physicians and patients, clarifying questions about the rare disease in question, its development, monitoring, complications, recommended follow-up examinations, and case discussions during the visit of the coordinator and/or at a distance. As the coordinators travel great distances for visits, often there is a request of lectures to physicians and patients from that region so that a larger number of people can solve their questions.

The 2009 Annual Report of the Fabry Registry (Sims et al. 2009) pointed out the importance of the characteristics of voluntary collection of observational data; patients are not randomized into groups "with treatment" and "no treatment," and in general the treated patients constitute more severe cases, what generates inherent biases. This report recommends that these and other factors must be considered when the registration data are evaluated.

The registry programs have developed guidelines for patients, periodical clinical and laboratory evaluation, but because registration is voluntary, the doctor accompanying the patient is who settles the kind of assessments and their frequency for each patient. This reflects the quality of data, especially in the long term.

Databases that combine features of protocol-oriented and practice-oriented information have as a goal to create a large and diverse source of prospective longitudinal patient data (Thadani 2006). The registries cited here are examples of such a database.

Data from the registries of the diseases herein described were supplied by the databases with the consent of the Advisory Boards, and the data were collected until October 2009.

Data Presentation and Discussion

The first registry of LSD was the Gaucher Registry, initiated in 1991, to fulfill a requirement of the FDA to monitor a larger number of patients with and without treatment.

The demographics data of the Gaucher Registry (Fig. 1) show a total of 5,661 patients, being 16% of this population

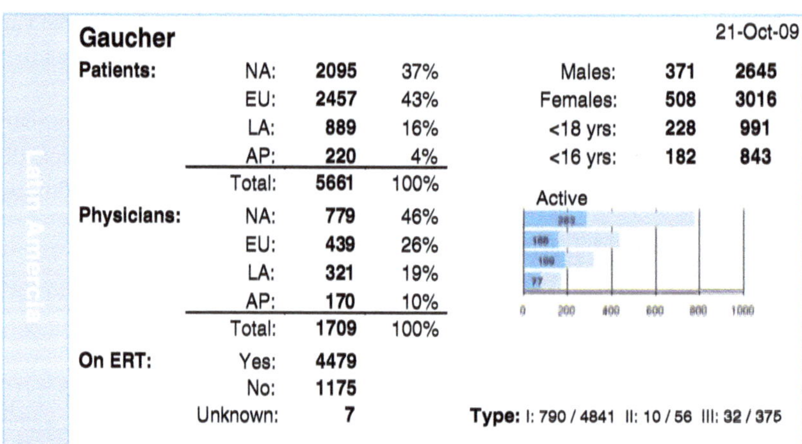

Fig. 1 Gaucher Registry overview

and 19% of all physicians participating in the program from Latin America. It is observed that the majority of patients have the type 1, the non-neuropathic form (92%). The largest proportion of patients receiving treatment live in Latin America (76%) in comparison with the rest of the world can probably be because most patients being diagnosed in Latin America are serious cases and not the oligosymptomatic. In the clinical practice, this information is important to give us the children's treatment because if we diagnosed early is because they are more severe.

In the Latin America 2009 Report of the International Collaborative Group of Gaucher Disease (ICGG) (unpublished data), it is observed that the majority of physicians participating in the Gaucher Registry in Latin America have 1–2 patients; about 63% and 14% have 3–4, just like in the USA and Europe (Fig. 2). In countries with large territories such as Brazil, this means a lot of work, as coordinators have to travel large distances to collect the data. Sometimes they are assisted by a trained person who enters information in the online programs.

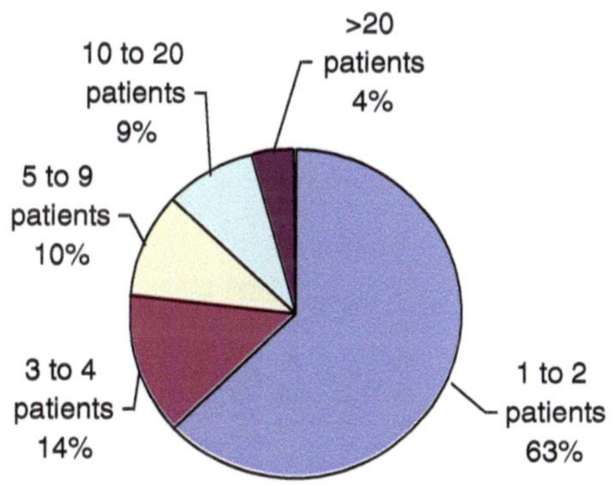

Fig. 2 Distribution of all registry patients per physician among patients with all disease types in Latin America

In the Latin American Annual Report (ICGG Gaucher Registry 2009), the number of hematologists involved in the treatment of patients with Gaucher disease in Latin America is almost three times higher compared to the rest of the world (unpublished data), and interestingly, the number of participating doctors without expertise specified is much greater in the rest of the world than in Latin America. We realized that in Latin America, Gaucher disease is yet best known among hematologists (Fig. 3).

The Fabry Registry began in 2001 and has now 3,318 patients; 20% of doctors and 281 patients are from Latin America (Fig. 4). Approximately 40% of patients in Latin America are being treated with β-agalsidase, while in the rest of the world, the percentage is 59%, probably due to the difficulty in reimbursement for treatment by the governments of Latin America, and perhaps associated with the fact that the disease has a phenotypic impact smaller than other lysosomal diseases that use the enzyme replacement therapy as treatment, in addition to affecting more adults than children, which can also reinforce the poor appeal for inclusion in a high-cost treatment program.

The total number of patients registered with MPS I is 857, being 126 in Latin America and accounting for 15% from the rest of the world. The percentage of doctors in Latin America is 18%. The percentage of patients receiving treatment is 80%, being the largest one, compared with 54% in North America and 66% in Europe (Fig. 5). It is hypothesized that most patients diagnosed in Latin America consist in the most severe cases, and therefore the rates of treatment are higher; the fact that the disease affects the children more than the adults also have a role in these results.

The Pompe Registry has 763 patients, 53 from Latin America representing 7% of the total, with 47 participating physicians, representing 13% (Fig. 6). About half of patients in Latin America are under 18 years of age, and in the rest of the world this percentage is 33%; so Latin America is diagnosing more children than adults, perhaps due to the more

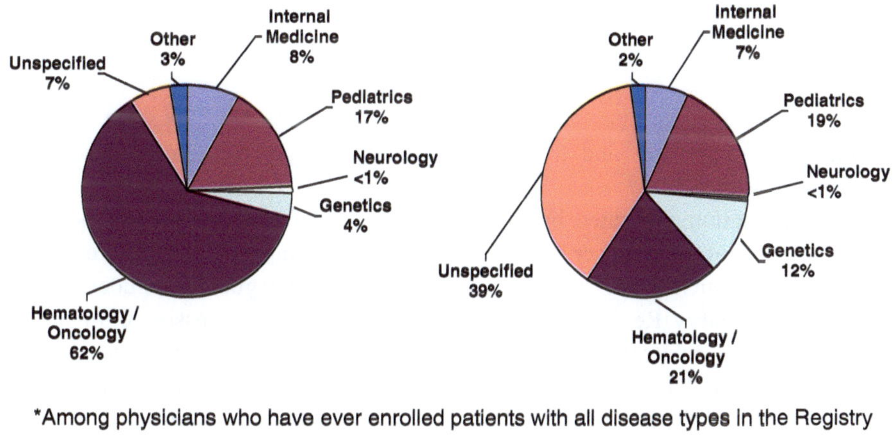

Fig. 3 Distribution of physician specialties (among physicians who have ever enrolled patients with all disease types in the registry) in Latin America and in the rest of the world

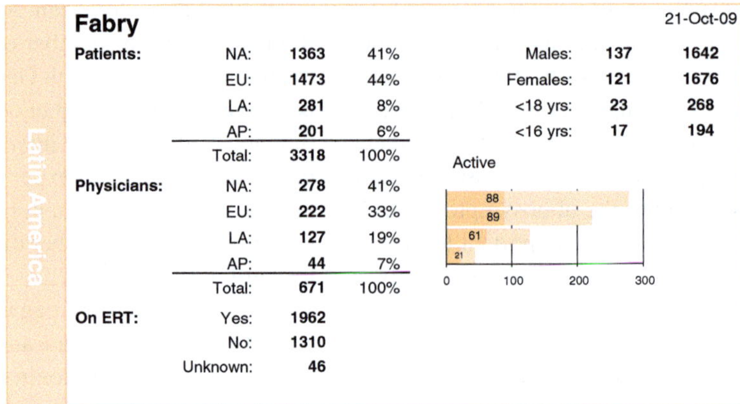

Fig. 4 Fabry Registry overview

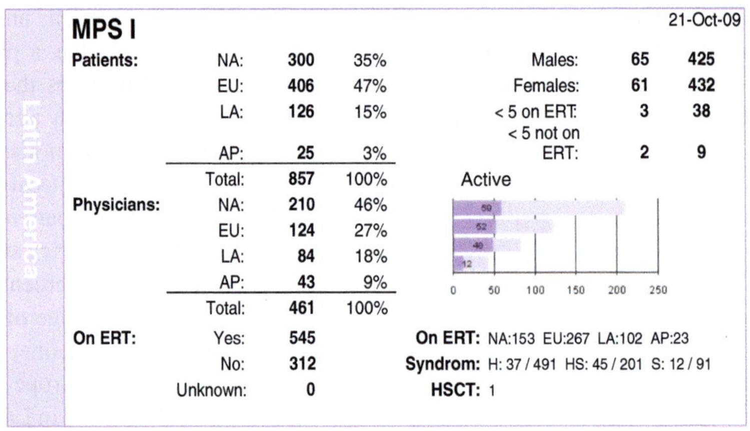

Fig. 5 MPS I Registry overview

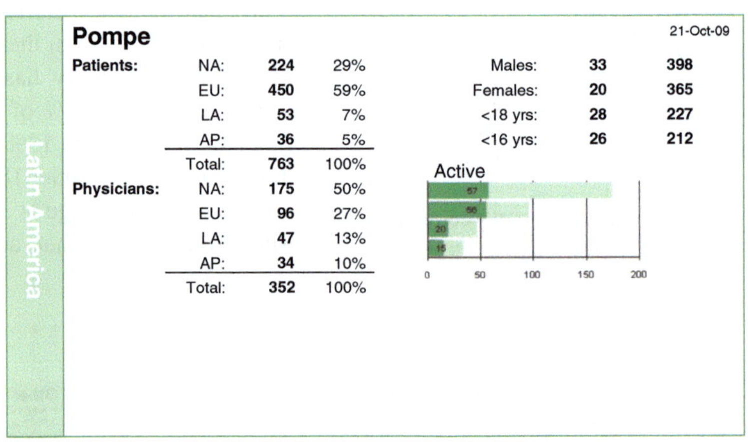

Fig. 6 Pompe Registry overview

exuberant clinical characteristics and to the more severe clinical manifestations.

The scientific production generated by the registries so far is 22 publications, 18 of the Gaucher Registry – with only one publication of the Gaucher Registry focusing on Latin America, seven of the Fabry Registry, two of the MPS I Registry, but with scarce publications focusing on the patient population of Latin America. The only publication (Sobreira et al. 2007) that focuses on the Latin American population, in this case specifically the Brazilian patients in the Gaucher Registry, aimed at the phenotypic and genotypic heterogeneity in type I Gaucher disease. When Brazil was compared with the rest-of-the-world patients, Sobreira and colleagues concluded that Brazilian patients may have a more aggressive clinical form of the disease than the rest of the world, what emphasizes the need for caution in making generalizations about Gaucher disease across demographic groups.

Final Considerations

Databases are very useful for doctors interested in the enzyme replacement therapy available for lysosomal diseases, and also for governments to anticipate the costs associated with the treatment and follow-up. In the long term, we can calculate the impact of therapeutic measures applied on the quality of life and survival of the patients. The disadvantage of these existing programs is its high cost, they are voluntary-based and observational, and it is difficult to achieve clinical and laboratory data at the appropriate frequency and quality to better understand the disease progression and its response to treatment. In Latin America is important to have information about our population, in comparison with USA and Europe as severity of the disease, response to ERT, and other characteristics of these rare diseases to indivudualize their treatments.

Acknowledgments The authors would like to thank Genzyme Corporation, Genzyme do Brasil, ICGG Gaucher Registry, Fabry Registry, MPS I Registry, Pompe Registry, to their support and the physicians and patients to their participation in the registries.

References

Sims K, Lee P, Wilcox W, Wanner C, Martins AM (2009) Fabry Registry Annual Report 2009. Published by Genzyme Corporation. http://www.fabry.org/fsig.psf/PDFs/PDFsR/$File/2009_RADAR.pdf

Sobreira E, Pires RF, Cizmarik M, Grabowski GA (2007) Phenotypic and genotypic heterogeneity in Gaucher disease type 1: a comparison between Brazil and the rest-of-the-world. Mol Genet Metab 90:81–86

Thadani R (2006) Formal trials versus observational studies. In: Metha A, Beck M, Sunder-Plassmann G (eds) Fabry disease, perspectives from 5 years of FOS. Oxford PharmaGenesis, Oxford. http://www.ncbi.nlm.nih.gov/bookshelf/

JIMD Reports
DOI: 10.1007/8904_2011_26

Received: 21 December 2010; Revised: 11 February 2011;
Accepted: 16 February 2011; Published online: 22 June
Communicated by: Jaak Jaeken

The Molecular Landscape of Phosphomannose Mutase Deficiency in Iberian Peninsula: Identification of 15 Population-Specific Mutations

B. Pérez, P. Briones, D. Quelhas, R. Artuch, A.I. Vega, E. Quintana, L. Gort, M.J. Ecay, G. Matthijs, M. Ugarte, and C. Pérez-Cerdá

Abstract PMM2-CDG is an autosomal recessive disorder and the most frequent form of congenital disorder of N-glycosylation, with more than 100 mutations identified to date. Sixty-six patients from 58 unrelated families were diagnosed as PMM2-CDG (CDG-Ia) based on clinical signs or because of a previous affected sibling. They all presented a type 1 serum transferrin isoform pattern, and, in most cases, the disease was confirmed by determining PMM2 activity in fibroblasts and/or lymphocytes. Residual PMM2 activity in fibroblasts ranged from not detectable to 60% of the mean controls. DNA and RNA were isolated from fresh blood or fibroblasts from patients to perform molecular studies of the *PMM2* gene, resulting in the identification of 30 different mutations, four of them newly reported here (p.Y102C, p.T118S, p.P184T, and p.D209G). From these 30 mutations, 15 have only been identified among Iberian PMM2-CDG patients. As in other Caucasian populations, p.R141H was the most frequent mutation (24 alleles, prevalence 20.6%), but less than in other European series in which this mutation represents 35–43% of the disease alleles. The next frequent mutations were p.D65Y (12 alleles, prevalence 10.3%) and p.T237M (9 alleles, prevalence 7.6%), while p.F119L and p.E139K, the most frequent changes in Scandinavian and French populations, respectively, were not found in our patients. The most common genotype was [p.R141H]+[p.T237M], and four homozygous patients for p.Y64C, p.D65Y, p.P113L, and p.T237M were detected. The broad mutational spectrum and the diversity of phenotypes found in the Iberian populations hamper genotype–phenotype correlation.

Keywords CDG · CDG-Ia · PMM2-CDG · Congenital defects of glycosylation · Mutations

B. Pérez, A.I. Vega, M.J. Ecay, M. Ugarte and C. Pérez-Cerdá (✉)
Centro de Diagnóstico de Enfermedades Moleculares, CBM-SO, Facultad de Ciencias, Módulo 10, Universidad Autónoma de Madrid, Cantoblanco, 28049 Madrid, Spain
and
CIBER de Enfermedades Raras (CIBERER) U746, Madrid, Spain
e-mail: cpcerda@cbm.uam.es

P. Briones
Sección de Errores Congénitos del Metabolismo (IBC), Servicio de Bioquímica y Genética Molecular, Hospital Clínic, IDIBAPS, Barcelona, Spain
and
CIBER de Enfermedades Raras (CIBERER) U737, Barcelona, Spain
and
Consejo Superior de Investigaciones Científicas, Barcelona, Spain

D. Quelhas
Medical Genetics Center, Porto, Portugal

R. Artuch
Servicio de Bioquímica, Hospital San Joan de Deu, Barcelona, Spain
and
CIBER de Enfermedades Raras (CIBERER) U722, Barcelona, Spain

E. Quintana and L. Gort
Sección de Errores Congénitos del Metabolismo (IBC), Servicio de Bioquímica y Genética Molecular, Hospital Clínic, IDIBAPS, Barcelona, Spain
and
CIBER de Enfermedades Raras (CIBERER) U737, Barcelona, Spain

G. Matthijs
Center for Human Genetics, Leuven, Belgium

Introduction

PMM2-CDG (CDG-Ia; MIM 212065) is the most frequent congenital disorder of the N-glycosylation pathways. It is caused by a deficiency in the cytosolic enzyme phosphomannomutase (PMM 2; EC 5.4.2.8) that converts mannose-6-phosphate to mannose-1-phosphate, by means of transient phosphorylation of the catalytic site of the enzyme. Mannose-1-phosphate is precursor in the formation of guanosine diphosphate-mannose (GDP-Man), the main mannose donor for *N*-, *O*-, and *C*-mannosylglycans and glycosylphosphatidylinositol (GPI)-anchor biosynthesis (Jaeken et al. 1997). The *PMM2* gene is located on chromosome 16q13 and has 8 exons and a reading frame of 738 base pairs encoding for a protein of 246 amino acids. Up to 109 mutations (HGMD Professional®) have been identified along the *PMM2* gene (MIM 601785), yielding a broad range of enzyme deficiencies (Matthijs et al. 1997; Haeuptle and Hennet 2009). There are mutational hot spots in exons 5 and 8. The majority of mutations (81%) are missense changes affecting different amino acid residues; 10% are nonsense or single or multiple base pair deletions which lead to early stop codons or frameshifts and thus to truncated proteins. Ten splicing defects have also been described due either to disruption of conserved sequences at the exon–intron junctions or at the polypyrimidine tract or to an intronic change that activates a pseudoexon sequence or to a deletion mediated by an Alu retrotransposition (Schollen et al. 2007a; Vega et al. 2009). The most common mutations are p.R141H and p.F119L, representing nearly 88% of mutant alleles described to date, both are probably founder mutations in Northern European countries (Bjursell et al. 1998). Most PMM2-CDG patients are compound heterozygotes for two different mutations, and homozygosity for p.R141H or other severely inactivating mutations has never been reported (Matthijs et al. 1998; Kjaergaard et al. 1998). Functional studies of several mutations have been performed to investigate their effect on activity and stability of the protein leading to the conclusion that a genotype retaining some residual PMM2 catalytic activity is required for survival (Pirard et al. 1999; Kjaergaard et al. 1999). It is worth noting that none of the amino acids constituting the active catalytic center by extrapolation in the crystal structure of the PMM1 isoenzyme has been found to be mutated (Silvaggi et al. 2006). The clinical course of patients with PMM2-CDG usually progress through four stages (Grunewald 2009): the infantile multisystemic or "visceral" stage, the childhood ataxia and mental retardation stage, the teenage leg atrophy stage, and the adult hypogonadal stage. Any organ system can be affected and severity of symptoms can widely vary. The most common presentation in infancy is multisystemic with central nervous system involvement. Mortality is as high as 20% of patients with the most severe form of the disease.

In a previous paper (Briones et al. 2002), we reported the biochemical and molecular characteristics of 26 Spanish PMM2-CDG patients. In this work, we update and report the predominant clinical signs and the spectrum of *PMM2* mutations in 66 PMM2-CDG patients from Portuguese and Spanish families.

Patients and Methods

Patients

The 58 families included were unrelated and originated from different regions of Portugal and Spain. In two families, one of the parents was from England and another from Russia, respectively. All cases were suspected due to clinical symptoms or because of a previous affected sibling, and they were referred for study by pediatricians from different hospitals since 1995.

Methods

Analysis of serum transferrin was performed as previously described by %CDT, isoelectric focusing (IEF) (Pérez-Cerdá et al. 2008; Colome et al. 2000) or by high-pressure liquid chromatography (HPLC) or capillary zone electrophoresis (Quintana et al. 2009). PMM2 activity in lymphocytes and fibroblasts was determined essentially as described in Van Schaftingen and Jaeken 1995. Fibroblasts were cultured according to standard procedures in minimal essential medium (MEM) supplemented with 1% glutamine, 10% bovine calf serum, and antibiotics in humidified atmosphere containing 5% of CO_2.

Genetic analysis was performed using whole blood or fibroblast cell lines from patients as source of mRNA and gDNA. Total mRNA was isolated by Tripure Isolation Reagent (Roche Applied Sciences, Indianapolis) adhering to the manufacturer's protocol, followed by two-step RT-PCR using oligodT as primer and using the SuperScript III First-Strand enzyme (Invitrogen, Carlsbad, CA).Genomic DNA was isolated using MagnaPure system, following the manufacturer's protocol (Roche Applied Science). The primers used for cDNA and gDNA amplifications were designed using the ENSEMBL database (http://www.ensembl.org/index.html; ENSG00000140650) and GenBank accession number NM_000303.2 as described in Vega et al. 2009. Amplifications of exons and flanking intronic sequences were performed using the FastStart kit (Roche Applied Sciences, Indianapolis). The PCR products were sequenced with the same primers used for amplification, using the BigDye Terminator v.3.1 mix (Applied Biosystems, Foster City, CA), following the manufacturers protocol. The sequenced products were then purified in Performa V3 96-Well Short Plates (EdgeBio, Gaithersburg, MD) and analyzed by capillary electrophoresis in an ABI Prism 3700 Genetic Analyzer (Applied Biosystems). DNA mutations numbering is based on cDNA reference sequence (GenBank accession number NM_000303.2) considering nucleotide +1 as the A of the ATG translation initiation codon and DNA reference sequence NT_010393.

Parental DNA was analyzed, when available, in heterozygous patients for allele assignment and in homozygous patients to rule out the presence of a genomic deletion.

Results

Sixty-six patients from 58 families were diagnosed as PMM2-CDG based on clinical signs, altered transferrin pattern, and deficient PMM2 activity in cells. All patients were of Caucasian origin. Collected clinical data of patients demonstrated that most of them presented with the "typical" clinical signs of a defect on protein glycosylation: cerebellar hypoplasia (82%), some degree of psychomotor retardation (76%), eye disturbances (54%), hypotonia (51%), ataxia (50%), liver dysfunction (36%), inverted nipples (36%), failure to thrive (36%), abnormal clotting factors (36%), dysmorphic features (29%), lipodystrophy (27%), and skeletal abnormalities (17%). Fifteen patients died, most of them in the first years of life, except for one in whom death occurred at age 13. Biochemically, they all presented an altered serum transferrin isoform pattern type 1, with increased amounts of asialo- and disialo-transferrin isoforms, compatible with a CDG type I diagnosis. PMM2-CDG was confirmed in most cases by determining PMM2 activity in fibroblasts and/or lymphocytes. Residual PMM2 activity in fibroblasts ranged from not detectable to as high as 60% of the mean of control values. In some patients in whom a high PMM2 activity value in fibroblasts was found, a lower enzyme activity in leukocytes ranging from 6% to 33% of the control values was demonstrated, confirming that the assay in leukocytes is more reliable for PMM2-CDG diagnosis. Molecular studies of the *PMM2* gene in RNA or DNA from 58 unrelated patients resulted in the identification of 30 mutations, four of them newly reported in this work (p.Y102C, p.T118S, p.P184T, and p.D209G). The results are shown in Table 1. As in other Caucasian populations, p.R141H is the most frequent mutation in the Spanish and Portuguese PMM2-CDG patients, accounting for 19.7% (17/86) and 21.8% (7/32) of the alleles, respectively. The second most frequent mutation in both populations was p.D65Y, accounting for 10% (12/118) of the disease alleles; but among the Spanish group of patients, the second most frequent was p.T237M that was found in 10.4% (9/86) of the disease alleles. Of the 30 identified variations, 13 have been reported only in our group of patients (p.M1V; p.Y64C; p.Y76C; IVS3-1G>C; p.E93A; p.Y102C; p.T118S; p.R123X; p.P184T; p.F207S; p.D209G; IVS7-9T>G; Del 28KB+exon8 Alu mediated). Forty-four different genotypes were detected; the most common one found in five Spanish families was [p.R141H]+[p.T237M]. Four homozygous patients for p.Y64C, p.D65Y, p.P113L, and p.T237M mutations were identified.

Discussion

PMM2-CDG is an autosomal recessive inherited disorder with an estimated incidence of 1:20,000 (Matthijs et al. 2000; Schollen et al. 2000). Due to the broad spectrum of clinical signs, including the very mild phenotypes recently described (Perez-Duenas et al. 2009; Grunewald 2009), the disease is probably still underdiagnosed. Of the patients of our series, 75% are currently alive and near a third of them are adults enjoying a good quality of life. This fact may be reflecting a lower prevalence of severe mutations in our population. Moreover, in contrast to other reported series where a low PMM2 activity (<15%) was found in patients' fibroblasts (Kjaergaard et al. 1998; Imtiaz et al. 2000), most of our patients did also show an enzyme activity below 25–30% of the control mean, but 15 patients presented a high activity up to 60% of control value. Some of these patients were functional hemyzygous for a mutation that retains in vitro a residual activity as high as 60% for p.C241S, 43% for p.P113L, and 48% for p.T237M or that affect the folding and/or stability of the enzyme (p.D65Y, p.V44A, and p.L32R) (Silvaggi et al. 2006; Vega et al. 2011), maybe contributing to the high residual activity found in vivo.

A large variety of mutations have been identified in our Iberian cohort of unrelated PMM2-CDG patients: 25 (83%) are missense mutations (Briones et al. 2002; Quelhas et al. 2007), four of them newly reported here (p.Y102C, p.T118S, p.P184T, and p.D209G); one nonsense (p.R123X) (Briones et al. 2002), three affecting the splicing of mRNA (IVS3+2T>C; IVS3-1G>C and IVS7-9T>C) (Briones et al. 2002; Vega et al. 2009), and a deletion mediated by an Alu retrotransposition displaying the complete loss of exon 8 (Schollen et al. 2007a). Of these 30 identified mutations, 13 have only been reported in our series of patients. The p.D65Y mutation was reported in a French study in a Portuguese patient, and a haplotypic association study confirmed its Iberian origin (Quelhas et al. 2007), and the p.V44A mutation was reported in an Ecuatorian patient probably of Spanish ancestors, adding these two mutations to our population-specific group of mutations. It is to note that mutations p.N216I and p.T226S were also previously reported in patients of Mediterranean origin (France, Italy, and Greece) (Matthijs et al. 2000). As expected, p.R141H was the most frequent mutation detected in our group of patients, but we have identified neither the severe mutation p.F119L, the second more frequent mutation among Scandinavian populations (Bjursell et al. 2000), nor p.E139K, the most prevalent change among French patients (Le Bizec

Table 1 Mutations in the *PMM2* gene in Iberian (Spanish and Portuguese) PMM2-CDG-Ia patients

Exon/ intron	Base change	Amino acid change	N° alleles, Spain (prevalence %)	N° alleles, Portugal (prevalence %)	Geographical origin[a]	Reference
1	c.1 A>G	p.M1V	1 (1.16)		ES	Perez-Duenas et al. (2009)
2	c.95TA>GC	p.L32R	1 (1.16)		AUS, CAN, ES, F, I	Matthijs (2000), Vega et al. (2009)
2	c.131 T>C	p.V44A	6 (6.97)		EC, ES	Matthijs et al. (1999), Briones et al. (2002)
3	c.191A>G	p.Y64C	3 (3.48)		ES	Briones et al. (2002)
3	c.193 G>T	p.D65Y	6 (6.97)	6 (20.0)	ES, F (Portuguese), PT	Matthijs et al. (1999), Briones et al. (2002), Quelhas et al. (2007)
3	c.227A>G	p.Y76C	1 (1.16)		ES	Matthijs et al. (2000), Briones et al. (2002)
In 3	c.255+2 T>C (IVS3+2 T>C)	Splice variant	1 (1.16)		D, ES	Matthijs et al. (1999), Briones et al. (2002)
In 3	c.256-1 G>C (IVS3-1 G>C)	p.Val60GlyfsX11	1 (1.16)		ES	Vega et al. (2009)
4	c.278A>C	p.E93A	1 (1.16)		ES	Briones et al. (2002)
4	**c. 305A>G**	**p.Y102C**	**1 (1.16)**		**ES**	**In this study**
4	c.338 C>T	p.P113L	6 (6.97)	1 (3.33)	AU, B, D, ES, F, NL, PL, PT, S, USA, JP	Matthijs et al. (1997), Briones et al. (2002)
5	**c.353 C>T**	**p.T118S**	**1(1.16)**		**ES**	**In this study**
5	c.367 C>T	p.R123X	1 (1.16)	1 (3.33)	ES, PT	Matthijs et al. (2000), Briones et al. (2002), Quelhas et al. (2007)
5	c.368 G>A	p.R123Q	5 (5.81)	3 (10.0)	AU, ES, F, I, NL, PT, S, USA	Matthijs et al. (1999), Briones et al. (2002), Quelhas et al. (2007)
5	c.385 G>A	p.V129M	3 (3.48)		CAN, ES, F, I, TK, USA	Matthijs et al. (1997), Briones et al. (2002)
5	c.422 G>A	p.R141H	17 (19.76)	7 (23.33)	AR, AU, AUS, B, CAN, CH, Czech, D, DK, EC, ES, F, I, IR, N, NL, PE, PT, S, UK, USA	Matthijs et al. (1997), Briones et al. (2002), Quelhas et al. (2007)
6	c.458 T>C	p.I153T	1 (1.16)	1 (3.33)	ES, F, PT, USA	Matthijs et al. (2000), Briones et al. (2002)
6	c.470 T>C	p.F157S	4 (4.65)	2 (6.66)	D, ES, F, I, PL, PT, USA	Matthijs et al. (1999), Briones et al. (2002), Quelhas et al. (2007)
6	c.484 C>T	p.R162W	2 (2.32)	3 (10.0)	B, ES, F, NL, PT, UK	Matthijs et al. (1997), Briones et al. (2002), Quelhas et al. (2007)
7	c.548 T>C	p.F183S	1 (1.16)		D, ES, S, UK	Matthijs et al. (2000), Briones et al. (2002)
7	**c.550 C>A**	**p.P184T**	**1 (1.16)**		**ES**	**In this study**
7	c.620 T>C	p.F207S	3 (3.48)		ES	Briones et al. (2002)
7	**c.626 A>G**	**p.D209G**	**1 (1.16)**		**ES**	**In this study**
In 7	c.640-9 T>G (IVS7-9 T>G)	p.Pro213_Gly214ins 23	2 (2.38)		ES	Vega et al. (2009)

8	c.647A>T	p.N216I	1 (1.16)		ES, Gr, I	Matthijs et al. (1997), Briones et al. (2002)
8	c.677 C>G	p.T226S	2 (2.32)	1 (3.33)	ES, F, PT	Matthijs et al. (2000), Briones et al. (2002), Quelhas et al. (2007)
8	c.691 G>A	p.V231M		3 (10.0)	AR, AU, B, CAN, D, F, I, PE, PT, S, UK, USA	Matthijs et al. (1997), Quelhas et al. (2007)
8	c.710 C>T	p.T237M	9 (10.465)		CAN, ES, F, I, IR, UK, USA	Matthijs et al. (1997), Briones et al. (2002)
8	c.722 G>C	p.C241S	4 (4.65)	1 (3.33)	B, ES, F, I, NL, PT, USA	Matthijs et al. (1999), Briones et al. (2002), Quelhas et al. (2007)
8	Alu retrotransposition-mediated deletion of 28 kb	Loss of exon 8	1 (1.16)		ES	Schollen et al. (2007b)

Mutations in bold are reported for the first time in this article. Changes p.T118S and p.P184T are located on the same allele

86 Spanish alleles (43 patients) and 30 Portuguese alleles (15 patients) were studied. All Spanish mutant alleles were identified and in one Portuguese allele (3.33%) the mutation remains unidentified

Aus Australia, *AU* Austria, *AR* Arab origin, *B* Belgium, *CAN* Canada, *CH* Switzerland, *Czech* Czech republic, *D* Germany, *DK* Denmark, *EC* Ecuador, *ES* Spain, *F* France, *GR* Greece, *I* Italy, *IR* Ireland, *JP* Japan, *N* Norway, *NL* the Netherlands, *PE* Peru, *PL* Poland, *PT* Portugal, *S* Sweden, *TK* Turkey, *UK* United kingdom, *USA* United States of America

[a]Information on geographical origin of mutations was obtained from database of Euroglycanet web: http://www.euroglycanet.org

et al. 2005). In summary, the different mutational spectrum with half of the mutations not reported till now in other populations, and the absence of the frequent mutations p. F119L and p.E139K, probably reflect the impact of Mediterranean migrations, as occurs for other metabolic inherited diseases such as phenylketonuria (Desviat et al. 1999; Mallolas et al. 1999) or galactosemia (Gort et al. 2006, 2009). It is noticeable that subtle differences were found between Spanish and Portuguese patients; so the third most frequent mutation (p.V231M) among European patients was only present in three Portuguese disease alleles, while p.T237M was only identified in Spaniards. It is difficult to know whether this observation is related to different population influences in Spain and Portugal or, more probably, to the short number of mutant alleles studied due to the rarity of the disease.

Most patients were functionally hemyzygous with one severe mutation and other that retains considerable residual activity, and no patients with two null mutations were detected, as it has ever been reported (Freeze and Westphal 2001; Le Bizec et al. 2005; Pirard et al. 1999), supporting again the hypotheses that the combination of two inactivating mutations is lethal (Schollen et al. 2000). Four homozygous patients were found, with mutations that probably affect the folding of the PMM2 monomer (Silvaggi et al. 2006). Those homozygosities result in mild phenotypes. Overall, 83% of the mutant chromosomes identified in our population are missense changes, most of them affecting the folding and stability of the PMM2 homodimer (misfolding or conformational disease) (Silvaggi et al. 2006; Pirard et al. 1999; Kjaergaard et al. 1999; Le Bizec et al. 2005; Vega et al. 2011), and thus probably contributing to the milder phenotypes found in comparison to those reported in other series (Erlandson et al. 2001; Imtiaz et al. 2000). Nevertheless, the highly variable phenotype described in PMM2-CDG, as in other monogenic diseases, is probably determined not only by the *PMM2* mutant alleles but also by the other genes modulating the effect on the final functional enzyme (Dipple and McCabe 2000).

Acknowledgments We thank patients, families, and pediatricians for their participation in this study. The work was supported by grants FIS PI07/118 and PI10/00455 from the Spanish Ministry of Health and the Sixth Framework program of the European Union (Euroglycanet: LSHM-CT-2005-512131).

References

Bjursell C, Wahlstrom J, Berg K, Stibler H, Kristiansson B, Matthijs G, Martinsson T (1998) Detailed mapping of the phosphomannomutase 2 (PMM2) gene and mutation detection enable improved analysis for Scandinavian CDG type I families. Eur J Hum Genet 6(6):603–611

Bjursell C, Erlandson A, Nordling M, Nilsson S, Wahlstrom J, Stibler H, Kristiansson B, Martinsson T (2000) PMM2 mutation spectrum, including 10 novel mutations, in a large CDG type 1A family material with a focus on Scandinavian families. Hum Mutat 16(5):395–400

Briones P, Vilaseca MA, Schollen E, Ferrer I, Maties M, Busquets C, Artuch R, Gort L, Marco M, van Schaftingen E et al (2002) Biochemical and molecular studies in 26 Spanish patients with congenital disorder of glycosylation type Ia. J Inherit Metab Dis 25(8):635–646

Colome C, Ferrer I, Artuch R, Vilaseca MA, Pineda M, Briones P (2000) Personal experience with the application of carbohydrate-deficient transferrin (CDT) assays to the detection of congenital disorders of glycosylation. Clin Chem Lab Med 38(10):965–969

Desviat LR, Perez B, Gamez A, Sanchez A, Garcia MJ, Martinez-Pardo M, Marchante C, Boveda D, Baldellou A, Arena J et al (1999) Genetic and phenotypic aspects of phenylalanine hydroxylase deficiency in Spain: molecular survey by regions. Eur J Hum Genet 7(3):386–392

Dipple KM, McCabe ER (2000) Modifier genes convert "simple" Mendelian disorders to complex traits. Mol Genet Metab 71(1–2):43–50

Erlandson A, Bjursell C, Stibler H, Kristiansson B, Wahlstrom J, Martinsson T (2001) Scandinavian CDG-Ia patients: genotype/phenotype correlation and geographic origin of founder mutations. Hum Genet 108(5):359–367

Freeze HH, Westphal V (2001) Balancing N-linked glycosylation to avoid disease. Biochimie 83(8):791–799

Gort L, Boleda MD, Tyfield L, Vilarinho L, Rivera I, Cardoso ML, Santos-Leite M, Giros M, Briones P (2006) Mutational spectrum of classical galactosaemia in Spain and Portugal. J Inherit Metab Dis 29(6):739–742

Gort L, Quintana E, Moliner S, Gonzalez-Quereda L, Lopez-Hernandez T, Briones P (2009) An update on the molecular analysis of classical galactosaemia patients diagnosed in Spain and Portugal: 7 new mutations in 17 new families. Med Clin (Barc) 132(18):709–711

Grunewald S (2009) The clinical spectrum of phosphomannomutase 2 deficiency (CDG-Ia). Biochim Biophys Acta 1792(9):827–834

Haeuptle MA, Hennet T (2009) Congenital disorders of glycosylation: an update on defects affecting the biosynthesis of dolichol-linked oligosaccharides. Hum Mutat 30(12):1628–1641

Imtiaz F, Worthington V, Champion M, Beesley C, Charlwood J, Clayton P, Keir G, Mian N, Winchester B (2000) Genotypes and phenotypes of patients in the UK with carbohydrate-deficient glycoprotein syndrome type 1. J Inherit Metab Dis 23(2):162–174

Jaeken J, Artigas J, Barone R, Fiumara A, de Koning TJ, Poll-The BT, de Rijk-van Andel JF, Hoffmann GF, Assmann B, Mayatepek E et al (1997) Phosphomannomutase deficiency is the main cause of carbohydrate-deficient glycoprotein syndrome with type I isoelectrofocusing pattern of serum sialotransferrins. J Inherit Metab Dis 20(3):447–449

Kjaergaard S, Skovby F, Schwartz M (1998) Absence of homozygosity for predominant mutations in PMM2 in Danish patients with carbohydrate-deficient glycoprotein syndrome type 1. Eur J Hum Genet 6(4):331–336

Kjaergaard S, Skovby F, Schwartz M (1999) Carbohydrate-deficient glycoprotein syndrome type 1A: expression and characterisation of wild type and mutant PMM2 in E. coli. Eur J Hum Genet 7(8):884–888

Le Bizec C, Vuillaumier-Barrot S, Barnier A, Dupre T, Durand G, Seta N (2005) A new insight into PMM2 mutations in the French population. Hum Mutat 25(5):504–505

Mallolas J, Vilaseca MA, Campistol J, Lambruschini N, Cambra FJ, Estivill X, Mila M (1999) Mutational spectrum of phenylalanine hydroxylase deficiency in the population resident in Catalonia: genotype-phenotype correlation. Hum Genet 105(5):468–473

Matthijs G (2000) Congenital disorders of glycosylation. Trends Biochem Sci 25(9):428

Matthijs G, Schollen E, Pardon E, Veiga-Da-Cunha M, Jaeken J, Cassiman JJ, Van Schaftingen E (1997) Mutations in PMM2, a phosphomannomutase gene on chromosome 16p13, in carbohydrate-deficient glycoprotein type I syndrome (Jaeken syndrome). Nat Genet 16(1):88–92

Matthijs G, Schollen E, Van Schaftingen E, Cassiman JJ, Jaeken J (1998) Lack of homozygotes for the most frequent disease allele in carbohydrate-deficient glycoprotein syndrome type 1A. Am J Hum Genet 62(3):542–550

Matthijs G, Schollen E, Heykants L, Grunewald S (1999) Phosphomannomutase deficiency: the molecular basis of the classical Jaeken syndrome (CDGS type Ia). Mol Genet Metab 68(2): 220–226

Matthijs G, Schollen E, Bjursell C, Erlandson A, Freeze H, Imtiaz F, Kjaergaard S, Martinsson T, Schwartz M, Seta N et al (2000) Mutations in PMM2 that cause congenital disorders of glycosylation, type Ia (CDG-Ia). Hum Mutat 16(5):386–394

Pérez-Cerdá C, Quelhas D, Vega AI, Ecay J, Vilarinho L, Ugarte M (2008) Screening using serum percentage of carbohydrate-deficient transferrin for congenital disorders of glycosylation in children with suspected metabolic disease. Clin Chem 54(1):93–100

Perez-Duenas B, Garcia-Cazorla A, Pineda M, Poo P, Campistol J, Cusi V, Schollen E, Matthijs G, Grunewald S, Briones P et al (2009) Long-term evolution of eight Spanish patients with CDG type Ia: typical and atypical manifestations. Eur J Paediatr Neurol 13 (5):444–451

Pirard M, Matthijs G, Heykants L, Schollen E, Grunewald S, Jaeken J, van Schaftingen E (1999) Effect of mutations found in carbohydrate-deficient glycoprotein syndrome type IA on the activity of phosphomannomutase 2. FEBS Lett 452(3):319–322

Quelhas D, Quental R, Vilarinho L, Amorim A, Azevedo L (2007) Congenital disorder of glycosylation type Ia: searching for the origin of common mutations in PMM2. Ann Hum Genet 71(Pt 3):348–353

Quintana E, Navarro-Sastre A, Hernandez-Perez JM, Garcia-Villoria J, Montero R, Artuch R, Ribes A, Briones P (2009) Screening for congenital disorders of glycosylation (CDG): transferrin HPLC versus isoelectric focusing (IEF). Clin Biochem 42(4–5):408–415

Schollen E, Kjaergaard S, Legius E, Schwartz M, Matthijs G (2000) Lack of Hardy-Weinberg equilibrium for the most prevalent PMM2 mutation in CDG-Ia (congenital disorders of glycosylation type Ia). Eur J Hum Genet 8(5):367–371

Schollen E, Keldermans L, Foulquier F, Briones P, Chabas A, Sanchez-Valverde F, Adamowicz M, Pronicka E, Wevers R, Matthijs G (2007a) Characterization of two unusual truncating PMM2 mutations in two CDG-Ia patients. Mol Genet Metab 90(4):408–413

Schollen E, Keldermans L, Foulquier F, Briones P, Chabas A, Sanchez-Valverde F, Adamowicz M, Pronicka E, Wevers R, Matthijs G (2007b) Characterization of two unusual truncating PMM2 mutations in two CDG-Ia patients. Mol Genet Metab 90(4):408–413

Silvaggi NR, Zhang C, Lu Z, Dai J, Dunaway-Mariano D, Allen KN (2006) The X-ray crystal structures of human alpha-phosphomannomutase 1 reveal the structural basis of congenital disorder of glycosylation type 1a. J Biol Chem 281(21):14918–14926

Van Schaftingen E, Jaeken J (1995) Phosphomannomutase deficiency is a cause of carbohydrate-deficient glycoprotein syndrome type I. FEBS Lett 377(3):318–320

Vega AI, Pérez-Cerdá C, Desviat LR, Matthijs G, Ugarte M, Perez B (2009) Functional analysis of three splicing mutations identified in the PMM2 gene: toward a new therapy for congenital disorder of glycosylation type Ia. Hum Mutat 30(5):795–803

Vega AI, Pérez-Cerdá C, Abia D, Gámez A, Briones P, Artuch A, Desviat LR, Ugarte M, and Pérez B (2011) Expression analysis revealing destabilizing mutations in phosphomannomutase 2 deficiency-congenital disorder of glycosylation. J Inherit Metab Dis 2011 May 4 [Epub ahead of print]

JIMD Reports
DOI: 10.1007/8904_2011_27

Received: 10 December 2010; Revised: 31 January 2011;
Accepted: 3 February 2011; Published online: 22 June
Communicated by: Garry Brown

Quantitative Analysis of mtDNA Content in Formalin-Fixed Paraffin-Embedded Muscle Tissue

Aida Font, Frederic Tort, Aleix Navarro-Sastre, Victòria Cusí, Judit García-Villoria, Paz Briones, and Antonia Ribes

Abstract Quantification of mitochondrial DNA (mtDNA) content is an essential tool for the diagnosis of mtDNA depletion syndrome (MDS). Samples collected and processed for anatomopathology studies represent a unique source of archived biological material. Thus, the possibility to study mtDNA copy number in these specimens would be a useful way to screen for MDS. In this study, we designed and validated the methodology to determine mtDNA content by quantitative real-time polymerase chain reaction (qRT-PCR) in formalin-fixed paraffin-embedded (FFPE) muscle tissue. We studied 14 frozen muscle biopsies and compared the results with a portion of the same biopsy embedded in paraffin. Our results showed a similar variability among frozen and FFPE muscle biopsies. Patients with MDS detected in frozen muscle were also confirmed in their corresponding FFPE samples, which validate the usefulness of this approach. We conclude that the analysis of mtDNA copy number in FFPE muscle tissue by qRT-PCR is a useful method for the molecular screening of patients suspected to have MDS when frozen biopsies are not available. Analysis of these samples would facilitate retrospective studies and diagnostic procedures.

A. Font
Sección de Errores Congénitos del Metabolismo, Servicio de Bioquímica y Genética Molecular, Hospital Clínic, Barcelona, Spain

F. Tort, A. Navarro-Sastre, and J. García-Villoria
Sección de Errores Congénitos del Metabolismo, Servicio de Bioquímica y Genética Molecular, Hospital Clínic, Barcelona, Spain
and
CIBER de Enfermedades Raras (CIBERER), Barcelona, Spain

V. Cusí
Servicio de Anatomía Patológica, Hospital San Joan de Déu, Barcelona, Spain

P. Briones
Sección de Errores Congénitos del Metabolismo, Servicio de Bioquímica y Genética Molecular, Hospital Clínic, Barcelona, Spain
and
CIBER de Enfermedades Raras (CIBERER), Barcelona, Spain
and
Consejo Superior de Investigaciones Científicas (CSIC), Barcelona, Spain

A. Ribes (✉)
Sección de Errores Congénitos del Metabolismo, Servicio de Bioquímica y Genética Molecular, Hospital Clínic, Barcelona, Spain
and
CIBER de Enfermedades Raras (CIBERER), Barcelona, Spain
and
Edificio Helios III, planta baja, C/Mejía Lequerica s/n, 08028 Barcelona, Spain
e-mail: aribes@clinic.ub.es

Introduction

Biogenesis and homeostasis of the mitochondria is tightly regulated, and requires the expression and coordination of both, nuclear and mitochondrial encoded proteins. Therefore, mitochondrial disorders are a group of complex dual genome diseases that can be caused by molecular defects in both nuclear or mitochondrial genomes including point mutations, deletions, duplications, and reduction in mitochondrial DNA (mtDNA) copy number, known as mtDNA depletion (Rötig and Poulton 2009).

mtDNA depletion syndromes (MDS) are a heterogeneous group of autosomal recessive disorders characterized by a reduction of the mtDNA content in a tissue-specific manner. They are caused by molecular defects in nuclear genes responsible for the biogenesis and maintenance of mtDNA integrity and usually affect different tissues and organs with high energetic demand, such as liver, skeletal muscle, and nervous system (Suomalainen and Isohanni 2010). Currently, MDS are divided into different syndromes caused by mutations in at least nine genes: myopathic form associated with mutations in *TK2* (OMIM # 609560); encephalomyopathic with renal tubulopathy form associated with *RRM2B* (OMIM #612075); encephalomyopathic with methylmalonic aciduria associated with mutations in *SUCLA2* and *SUCLG1* (OMIM #612073 and 245400); hepatocerebral form associated with mutations in *DGUOK*, *MPV17*, and *TWINKLE* (OMIM # 251880, 256810 and 271245); MNGIE syndrome associated with mutations in *TYMP* and *POLG* (OMIM # 603041 and 613662). Mutations in *POLG* are also associated with Alpers Syndrome (OMIM #203700) (Dimmock et al. 2010).

Quantification of mtDNA content is an essential tool for the diagnosis of MDS and is based on Southern blot or quantitative real-time polymerase chain reaction (qRT-PCR) analysis (Suomalainen and Isohanni 2010). Although DNA from fresh frozen tissues is the suitable sample for the diagnosis of MDS, the most widely used method to collect samples is formalin fixation followed by paraffin block inclusion. The feasibility of performing mtDNA quantification on formalin-fixed paraffin-embedded (FFPE) tissues would facilitate diagnostic procedures and allow a large number of retrospective studies.

Material and Methods

Biological Samples

We studied 14 frozen skeletal muscle biopsies and compared them with a portion of the same biopsy embedded in paraffin. The age distribution and number of controls were: 0–1 year (n = 12); 6 years (n = 1), and 50 years (n = 1). To validate the methodology, we analyzed four patients with known reduction of mtDNA copy number. Briefly:

Patient1: She was admitted to the hospital at 7 days of life with bradypnea. She progressively developed hypothermia, mydriasis, and edema. She died few hours after admission. Biochemical analysis showed hyperammonemia, increased excretion of lactate, and hypertransaminasemia; 90% of mtDNA depletion in muscle biopsy by qRT-PCR was detected, while respiratory chain activities in the same muscle biopsy were normal. Mutations in *DGUOK* and *MPV17* were excluded.

Patient 2: He was admitted to the hospital at 30 h of life with feeding refuse, irritability, dehydration, severe metabolic acidosis, hyperammonemia, hyperlactatemia, and altered hepatic enzymes. At 8 days of age, neurological deterioration, as well as renal and hepatic insufficiency, was evident. He presented two further episodes of metabolic acidosis and liver involvement. He died of hepatic failure at 2 months of age; mtDNA content in muscle biopsy, determined by southern blot, showed 89% mtDNA depletion. Mitochondrial respiratory chain activities (complex I–IV) were low. Genetic studies excluded mutations in *DGUOK* and *RRM2B*.

Patient 3: She presented at 5 days of age with axial hypotonia, poor eye contact, and sweating. Biochemical studies revealed an increased excretion of lactate and methylmalonate. Mitochondrial respiratory chain activities (complex I–IV) showed a slight reduction of all the complexes; mtDNA content, determined by qRT-PCR, showed 87% mtDNA depletion in muscle biopsy. A new mutation in *SUCLA2* was identified, which has been predicted to be disease causing according to Polyphen database (unpublished results and currently under study in our laboratory).

Patient 4: She is the third daughter of a family with two other affected siblings. Few hours after birth, she presented a severe hepatic insufficiency, metabolic acidosis, hypoglycemia, hyperammonaemia, and high lactate. Mitochondrial respiratory chain activities were normal (complex I–IV); mtDNA content, determined by qRT-PCR in muscle biopsy, showed 80% depletion. Genetic analysis of *DGUOK* identified the previously described p.H226R mutation (Dimmock et al. 2008) in homozygosity.

Informed consent has been obtained for all the samples used.

Experimental Design and Sample Preparation

To minimize the natural variability of mtDNA content within tissue samples, we prepared and processed separately two adjacent portions from each frozen tissue. DNA was directly extracted from one of the portions, while the other one was subjected to formalin fixation and paraffin block inclusion. Briefly, before DNA isolation, paraffin sections were dewaxed in xylene at room temperature and subjected to a series of alcohol washing steps, following standard procedures (Gilbert et al. 2007). Then, the samples were air-dried and DNA was isolated using Qiamp DNA mini kit (Qiagen, GmbH, Germany).

Quantification of mtDNA Content

Analysis of mtDNA copy number was performed by qRT-PCR using Taqman technology in a Step One plus real-time PCR system (Applied Biosystems, Foster City, CA, USA) (Table 1). This method is based on the amplification of the *12S rRNA* (mtDNA target gene) and the *RNaseP* (endogenous nuclear control gene) to normalize the DNA content in each sample. PCRs were carried out in triplicate using 15 ng of DNA from both frozen and FFPE samples. Levels of mtDNA were relatively quantified by evaluating Ct values using the comparative Ct (ΔΔCt) method. Briefly, mtDNA content was calculated from the difference of the delta Ct of each sample (Ct*12SrRNA*−Ct*RNaseP*) compared with the reference value obtained as the average of delta Ct from the controls studied. Final values were expressed in relative units and were obtained through the equation $2^{-\Delta\Delta Ct}$

Table 1 Oligonucleotides and reagents used for qRT-PCR

Gene	Primers	Probes	Fluorescent product
12S rRNA[a]	CCACGGGAAACAGCAGTGAT CTATTGACTTGGGTTAATCGTGTGA	(6FAM)TGCCAGCCACCGCG(BHQ1)	6FAM/BlackHole Quencher
RNAseP[b]	Taqman RNase P Control Reagents Kit. Part Number 4316844		VIC-TAMRA

[a]Oligonucleotides are located at positions 158 and 280 according to *12SrRNA* sequence NC_012920; PCR amplification yields a 122 bp product
[b]According to the manufacturer, Taqman technology uses amplicons around 100 bp in length

where: $\Delta\Delta Ct = (Ct\textit{12SrRNA}\text{ sample} - Ct\textit{RNaseP}\text{ sample}) - (\text{average }\Delta Ct\text{ controls})$.

Statistical Analysis

Wilcoxon signed-rank (nonparametric two-related sample test) and Spearman correlation test were performed using SPSS16.0 statistical software.

Results and Discussion

FFPE specimens represent an important and unique source of archived biological material. According to the literature, molecular studies have been successfully performed in this type of samples (Gnanapragasam 2009; Yu et al. 2008). However, as far as we know, the reliability of qRT-PCR technology for mtDNA copy number determination in FFPE tissues remains still to be elucidated. Moreover, the heterogeneous clinical spectrum of MDS makes the mtDNA copy number study an essential tool before undertaking specific gene sequencing analysis (Bai et al. 2004; Dimmock et al. 2010). Thus, the possibility to detect mtDNA content in FFPE specimens would represent a useful way to screen for MDS. Our aim was to optimize the methodology to determine mtDNA content by qRT-PCR in FFPE muscle tissue.

Limitations of molecular biology studies in FFPE tissues have been widely reported to be due to the high fragmentation of nucleic acids. We visualized the integrity of the DNA from frozen and FFPE samples in agarose gels and results showed that the DNA was highly fragmented in comparison with that extracted from frozen samples (Fig. 1a). Despite the concentration of DNA is lower in FFPE than in frozen muscle samples, the purity of both is almost identical and suitable for molecular studies (data not shown). According to the literature, the median DNA fragment length in FFPE samples is around 400 bp and mainly depends on the tissue type and the methodology of fixation (Gilbert et al. 2007; Lehmann and Kreipe 2001). To determine the extent of the degradation of the DNA in FFPE samples, we performed conventional PCR. Two nuclear and one mtDNA fragments of around 1,200, 700 and 500 bp, respectively, were successfully amplified (Fig. 1b). The lower intensity of the 1,200 bp fragment in the FFPE samples is probably due to DNA degradation during formalin fixation and sample processing. Because the qRT-PCR amplicons had around 100 bp, we considered that the DNA obtained from FFPE samples was suitable for mtDNA determination. However, FFPE samples showed higher Ct values in comparison with their corresponding frozen tissue (Fig.1c). This phenomenon could be explained by the fact that the DNA degradation in FFPE samples might reduce the number of intact amplicons, although the starting DNA amounts in the qRT-PCR were the same in both FFPE and frozen muscle. Because the delta Ct values obtained in both kind of samples were slightly different (mean; −10.54 and −11.72, respectively) the two groups were analyzed separately.

A similar variability in the mtDNA content between both groups has been observed, reference range: 0.57–2.90 and 0.41–2.02, in frozen and FFPE samples, respectively (Fig. 2a). Wilcoxon signed-rank test showed no statistically significant differences (p-value = 0.95). To assess the level of agreement of paired samples, the Spearman's correlation test was applied and showed a good correlation within specimens ($r = 0.864$; $p < 0.01$). Altogether these data demonstrate that the mtDNA content in frozen and FFPE muscle samples was comparable. Despite the variability observed within these two groups, the comparative study showed similar behavior for each sample (Fig. 2b–c). Nevertheless, as has been reported previously (Bai et al. 2004; Morten et al. 2007; Dimmock et al. 2010), the observed variability could be explained by the differences in the age of control individuals. Interestingly, one of the control samples (C11), showing the highest mtDNA content, belongs to an adult while the remaining samples are from pediatric controls (Fig. 2c).

The four patients (P1–P4) with MDS detected in frozen muscle were clearly confirmed in their corresponding FFPE samples when compared with age-matched controls (Fig. 2d), which validates the usefulness of this material. Thus, despite the degradation of the DNA, mtDNA depletion can be detected in FFPE specimens in a similar manner to that performed in frozen tissues. However, the relatively small number of specimens analyzed in this study should

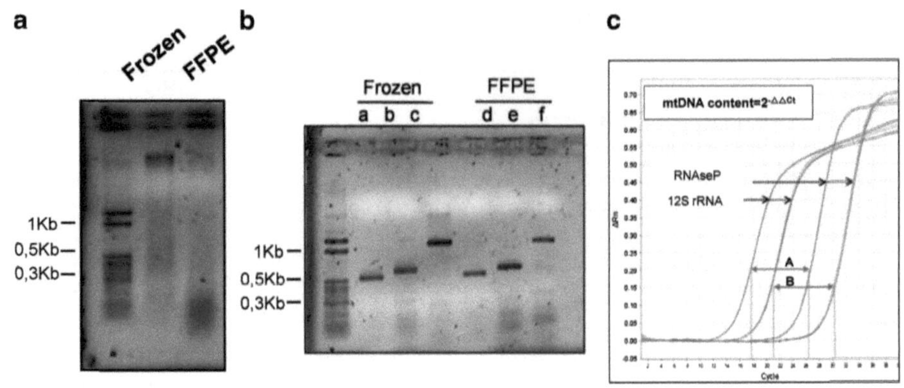

Fig. 1 Integrity of the DNA obtained from frozen and FFPE muscle biopsy. (**a**) Total genomic DNA was visualized in 1% agarose gel. MW, molecular weight marker (**b**) DNA products from frozen and FFPE muscle biopsy obtained by PCR amplification. MW, molecular weight marker; lanes *b*, *c*, *e*, and *f* correspond to nuclear DNA fragments; lanes *a* and *d* correspond to a mtDNA fragment. (**c**) qRT-PCR amplification plot in frozen (*A*) and FFPE (*B*) samples

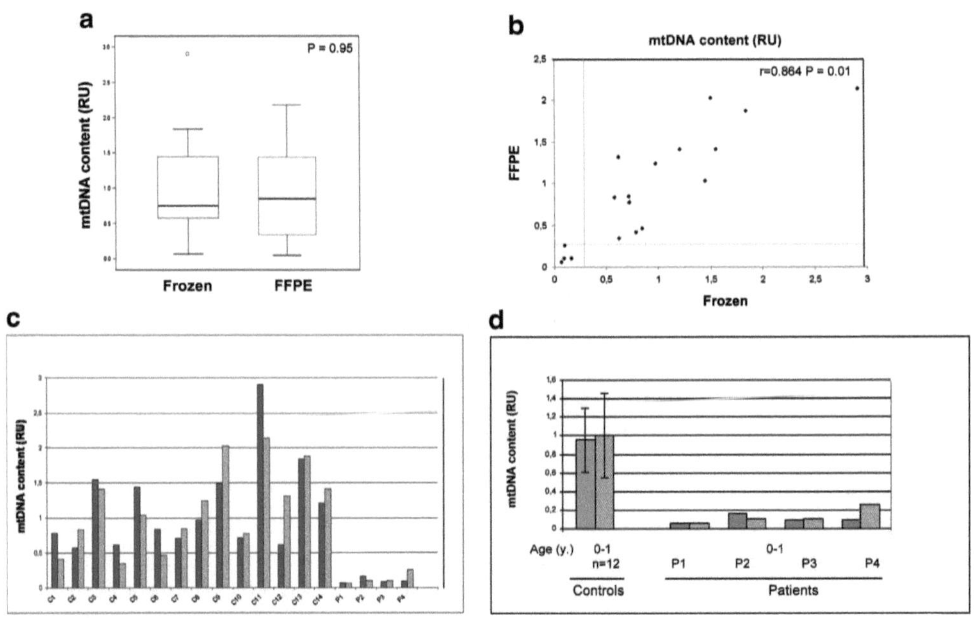

Fig. 2 Comparative analysis of mtDNA content in frozen and FFPE muscle biopsies. (**a**) Similar levels of mtDNA were found in frozen and FFPE samples as determined by the Wilcoxon ranked test. (**b**) Spearman test showed statistically significant correlation between FFPE and frozen paired samples. Dotted line separates tissues with confirmed mtDNA depletion and delimits the threshold for mtDNA depletion (30% of mtDNA content in control samples). (**c**) mtDNA content in frozen and FFPE paired samples. C1–C14 correspond to control samples and P1–P4 to patients with previously confirmed mtDNA depletion. Results are shown in relative units (RU). Gray bars indicate frozen tissue, striped bars indicate FFPE samples. (**d**) Comparison of the mtDNA content of the four patients (P1–P4) with age-matched controls. Results are shown in relative units (RU). Gray bars indicate frozen tissue, striped bars indicate FFPE samples. Error lines represent the mean ± standard deviation

be supported by further studies in larger series of cases to fully confirm our observations.

It has also been reported that many variables can influence the validity and reliability of molecular biology studies in FFPE tissues, such as tissue amount, length, storage, and fixation conditions (Gnanapragasam 2009; Gilbert et al. 2007). Therefore, the above-mentioned aspects should also be considered when analyzing mtDNA content in FFPE tissues.

In conclusion, analysis of mtDNA copy number in FFPE by qRT-PCR is a useful method for molecular screening of patients suspected to have MDS when frozen biopsies are not available. Analysis of these samples, which are a rich source of biological material, would facilitate retrospective studies and diagnostic procedures.

Acknowledgements This research was supported in part by the Spanish Ministerio de Sanidad Grant PI08/90348. CIBERER, is an initiative of Instituto de Salud Carlos III.

References

Bai RK, Perng CL, Hsu CH, Wong LJ (2004) Quantitative PCR analysis of mitochondrial DNA content in patients with mitochondrial disease. Ann NY Acad Sci 1011:304–309

Dimmock D, Zhang Q, Dionisi-Vici C et al (2008) Clinical and molecular features of mitochondrial DNA depletion due to mutations in deoxyguanosine kinase. Hum Mutat 29(2):330–331

Dimmock D, Tang L, Schmitt E, Wong L (2010) Quantitative evaluation of the mitochondrial DNA depletion syndrome. Clin Chem 56:1119–1127

Gilbert M, Haselkorn T, Bunce M et al (2007) The isolation of nucleic acids from fixed, paraffin-embedded tissues-which methods are useful and when? PLoS ONE 20(2):e537

Gnanapragasam VJ (2009) Unlocking the molecular archive: the emerging use of formalin-fixed paraffin-embedded tissue for biomarker research in urological cancer. BJU Int 105:274–278

Lehmann U, Kreipe H (2001) Real-Time PCR analysis of DNA and RNA extracted from formalin-fixed and paraffin-embedded biopsies. Methods 25:409–418

Morten KJ, Ashley N, Wijburg F et al (2007) Liver mtDNA content increases during development: a comparison of methods and the importance of age- and tissue-specific controls for the diagnosis of mtDNA depletion. Mitochondrion 7(6):386–395

Rötig A, Poulton J (2009) Genetic causes of mitochondrial DNA depletion in humans. Biochim Biophys Acta 1792:1103–1108

Suomalainen A, Isohanni P (2010) Mitochondrial DNA depletion syndromes-many genes, common mechanisms. Neuromuscul Disord 20:429–437

Yu J, Miller R, Zhang W et al (2008) Copy-number analysis of topoisomerase and thymidylate synthase genes in frozen and FFPE DNAs of colorectal cancers. Pharmacogenomics 9: 1459–1466

JIMD Reports
DOI: 10.1007/8904_2011_28

Received: 27 August 2010; Revised: 9 October 2010;
Accepted: 19 October 2010; Published online: 25 June
Communicated by: Bridget Wilcken

Relevance of Expanded Neonatal Screening of Medium-Chain Acyl Co-A Dehydrogenase Deficiency: Outcome of a Decade in Galicia (Spain)

M.L. Couce, D.E. Castiñeiras, J.D. Moure, J.A. Cocho, P. Sánchez-Pintos, J. García-Villoria, D. Quelhas, N. Gregersen, B.S. Andresen, A. Ribes, and J.M. Fraga

Abstract Neonatal screening of medium-chain acyl-CoA dehydrogenase deficiency (MCADD) is of major importance due to the significant morbidity and mortality in undiagnosed patients. MCADD screening has been performed routinely in Galicia since July 2000, and until now 199,943 newborns have been screened. We identified 11 cases of MCADD, which gives an incidence of 1/18,134. During this period, no false negative screens have been detected. At diagnosis, all identified newborns were asymptomatic. Our data showed that octanoylcarnitine (C8) and C8/C10 ratio are the best markers for screening of MCADD. C8 was increased in all patients and C8/C10 was increased in all but one patient.

The common mutation, c.985A>G, was found in homozygosity in seven newborns and in compound heterozygosity in three, while one patient did not carry the common mutation at all. In addition, two novel mutations c.245G>C (p.W82S) and c.542A>G (p.D181G) were identified. Ten of the 11 identified newborns did not experience any episodes of decompensation. The patient with the highest level of medium chain acylcarnitines at diagnosis, who was homozygous for the c.985A>G mutation, died at the age of 2 years due to a severe infection.

This is the first report of the results from neonatal screening for MCADD in Spain. Our data provide further evidence of the benefits of MCADD screening and contribute to better understanding of this disease.

M.L. Couce (✉), D.E. Castiñeiras, J.D. Moure, J.A. Cocho, P. Sánchez-Pintos, and J.M. Fraga
Unidad de Diagnóstico y Tratamiento de Enfermedades Metabólicas, Departamento de Pediatría, Hospital Clínico Universitario, Universidad de Santiago, A Choupana s/n. 15706, Santiago de Compostela, Spain
e-mail: maria.luz.couce.pico@sergas.es

J. García-Villoria, and A. Ribes
Sección de Errores Congénitos del Metabolismo (IBC), Servicio de Bioquímica y Genética Molecular, Hospital Clínic y Centro de Investigación Biomédica en Red de Enfermedades Raras (CIBERER), Barcelona, Spain

D. Quelhas
Centro de Genética Médica Jacinto Magalhaes, Porto, Portugal

N. Gregersen
Research Unit for Molecular Medicine, Aarhus University Hospital, Skejby, 8200 Aarhus N, Denmark

B.S. Andresen
Research Unit for Molecular Medicine, Aarhus University Hospital, Skejby, 8200 Aarhus N, Denmark
and
Department of Biochemistry and Molecular Biology, University of Southern Denmark, 5230 Odense, Denmark

Introduction

Medium-chain acyl-CoA dehydrogenase deficiency (MCADD, OMIM #201450) is the most commonly inherited defect of the mitochondrial fatty acid oxidation pathway, with significant morbidity and mortality in undiagnosed patients. The estimated incidence varies between 1/5,000 and 1/68,560 according to the data reported in several studies (Yokota et al. 1991; Andresen et al. 2001; Tran et al. 2007; Wilcken et al. 2007).

MCADD is a recessively inherited disorder and homozygosity for the prevalent c.985A>G (K329E) mutation accounts for up to 80–90% of the detected cases (Yokota et al. 1991; Gregersen et al. 1991; Nennstiel-Ratzel et al. 2005; Giroux et al. 2007). To date, 81 *ACADM* mutations have been described in the Human Genome Mutation database http://www.hgmd.cf.ac.uk, but a clear genotype–phenotype correlation has not been established, and there is a wide phenotypic variability even within the same family (Yokota et al. 1991; Andresen et al. 1997; Lehotay et al. 2004; Waddell et al. 2006; Hsu et al. 2008).

Patients with MCADD have the inability to completely metabolize long chain fatty acids released from adipose tissues during catabolism that usually results in the sequestration of CoA and other biochemical and physiological features of the disorder.

Clinical symptoms are diverse ranging from asymptomatic to hypoketotic hypoglycemic episodes and even sudden death in some patients (Andresen et al. 1993, 2001; Waddell et al. 2006). Symptoms could occur at any age and are typically precipitated by stress in situations of high energy demand such as a prolonged period of fasting or an intercurrent illness, particularly between 3 and 24 months of age. Correct dietary management avoids or minimizes the number of metabolic decompensation episodes. Therefore, early identification of MCADD deficient individuals is important and newborn screening (NBS) by tandem mass spectrometry (MS/MS) for MCADD (Millington et al. 1990) has been initiated in numerous countries over the last decade. In Galicia (northwest Spain), NBS by MS/MS was started in July 2000. The main objective of this study was to evaluate the diagnostic results as well as the outcome of the MCAD deficient individuals after 10 years of screening.

Material and Methods

Analytical Methods

Until 2001, blood spots and urine-impregnated filter paper sampled between the 5th and 8th day of life were collected for neonatal screening in Galicia. In 2002, this recommendation changed and sampling on the third day of life, after 48 h of milk intake, was established.

Medium-chain acylcarnitines C8, C6, C10, C10:1, and the ratios C8/C2, and C8/C10 in blood spots were studied by the usual method of butylation and analysis by MS/MS (Chace et al. 2001). MCADD was suspected in those newborn who presented an elevation higher than the 99.9th percentile of the specific acylcarnitines (C6, C8, C10, C10:1). Cutoff values were obtained through screening of the population (n=199,943 samples): C8 > 0.52 μM, C6 > 0.43 μM, C10 > 0.50 μM, C10:1 > 0.33 μM, C8/C2 > 0.02, and C8/C10 > 1.85

Plasma Free Fatty acids (FFAs) were analyzed as previously described (Martínez et al. 1997) by specific methylation with acetyl chloride/methanol, in one step extraction and derivatization.

Analysis of urinary organic acids was performed by gas chromatography mass spectrometry (GC/MS) by the usual method of solvent extraction and trimethylsilyl derivatization (Tanaka et al. 1980).

DNA was isolated and sequenced by standard procedures from blood samples of all the patients and their parents, except one case conceived by in vitro fertilization (IVF) with oocyte donation (patient 9, Table 1), whose biological mother was not studied for obvious reasons.

Patients at Diagnosis

During the period of study, 199,943 newborn samples were tested and 11 cases from ten families were diagnosed (patients 3 and 4 were brothers). Eight cases were Caucasian and three were of Gypsy ethnicity.

At diagnosis, the following parameters were evaluated: age, sex, presence or absence of clinical symptoms, medium chain acylcarnitines and their ratios, plasma FFA, and urinary acylglycines and organic acids. The diagnosis was confirmed by mutation analysis of the *ACADM* gene (OMIM#607008). Clinical course was subsequently monitored.

Patients' Follow-up

Follow-up of all the patients, except one, has been done in our hospital.

Anthropometric evaluation: body weight, body length, and head circumference were measured and expressed as percentiles with respect to the reference population.

Evaluation of cognitive and psychomotor development: we reassessed Psychomotor Development Index (PDI) or Intellectual Quotient (IQ) of survivors using the Brunet Lézine Scale in infants, the Mc.Carthy Scales of Psychomotor Skills (MSCA) in preschool children and Wechler Intelligence Scale for Children Revised (WISC-R) in children older than 6 years. The overall index score of PDI or IQ is considered in the normal range when it is above 85.

Dietary management: the patients received a normal age diet allowing normal growth and development and avoiding prolonged fasting periods and lipolisis by high intake of slow absorption carbohydrates. We used a computer program of dietary calculation developed by ourselves that monitors fat intake and energy (www.odimet.es). Biochemical follow-up includes measurement of plasma-free carnitine at each visit and annually determination of general parameters including transaminases. Carnitine supplement is prescribed if the level is below 12 μM.

Informed consent of the patients' parents was obtained. The study was approved by the Ethics Committee of our hospital.

Results

Since the introduction of the expanded NBS in Galicia, we identified 11 cases with MCADD, which represents an incidence of 1/18,176 newborns. All of them, eight males and three females, were born at term with normal birth weight. The medium age of sample collection for screening was 7.4 days

Table 1 Summary of laboratory investigations, mutations at MCADD diagnosis and evolution.

	Acylcarnitines - mol/L (99.9th percentile)					Free fatty acids -mol/L (control values)			Mutation*	Follow-up (years)	PDI/IQ	Present status
Patients	C8 (<0.52)	C10 (<0.5)	C10:1 (<0.33)	C8/C10 (<1.85)	C8/C2 (<0.02)	C8:0 (≤1-8)	C10:0 (4-9)	C10:1 (≤0.4-1)				
1	1.31	0.20	0.27	6.62	0.12	80	22	10	p.K329E/p.K329E	6	104	FS
2	1.23	0.17	0.20	7.33	0.11	35	10	16	p.K329E/p.K329E	6	98	FS
3	3.06	0.38	0.66	7.94	0.34	200	12	28	p.K329E/p.K329E	5	87	FS
4	5.82	0.56	0.47	10.3	0.67	93	25	38	p.K329E/p.K329E	1	nd	Exitus
5	1.9	0.17	0.22	11.0	0.09	76	88	4.2	p.K329E/p.K329E	3	105	FS
6	9.4	0.71	0.97	13.2	0.74	nd	nd	nd	p.K329E/p.K329E	1	nd	FS
7	4.57	0.40	0.55	11.4	0.55	nd	nd	nd	p.K329E/p.K329E	1	nd	FS
8	0.84	0.50	0.33	1.64	0.07	6.8	6.8	4.2	p.K329E/p.Y67H	6	128	FS
9	1.72	0.22	0.61	7.53	0.22	120	28	6.1	p.K329E/p.W82S	6	142	FS
10	1.26	0.51	0.39	2.45	0.12	nd	nd	nd	p.K329E/p.I416T	0.6	nd	FS
11	5.19	0.81	0.56	6.38	0.6	nd	nd	nd	p.G267R/p.D181G	0.8	103	FS

FS: Free of Symptoms; nd= not done

PDI/IQ: Psychomotor Development Index/ Intellectual Quotient

*All mutations were named according to the precursor enzyme.

(range: 4–23). At diagnosis, all the patients were asymptomatic and presented with increased levels of medium chain acylcarnitines. As shown in Table 1, they exhibited a marked elevation of octanoylcarnitine (C8) with a median value of 3.3 μmol/L. The highest values were observed in patients 4 and 6 and the lowest was observed in patient 8. The C8/C10 ratio showed a clear elevation (median value: 7.79 μmol/L) reaching the highest value in patient 6. In our hands, C8 has demonstrated to be the best parameter for the diagnosis of MCADD; C8/C10 ratio was also good but it was normal in one patient.

During the period of study, two individuals only heterozygous for the common c.985A>G mutation were identified, presenting a slight elevation of octanoylcarnitine 0.54 μmol/L, and 0.56 μmol/L, respectively. Other mutations were excluded by sequencing all the exons and intron boundaries of the *ACADM* gene. To our knowledge, no false negative cases have been identified in our region.

Plasma FFA were studied in seven patients (Table 1), the profile showed elevations of C8:0, median value 87.4 μmol/L (Reference < 8), C10:0, median value 27.4 μmol/L (Reference < 9) and C10:1, median value 15.13 μmol/L (Reference < 1). Patient 3 presented the highest C8:0 (200 μmol/L), patient 4 the highest C10:1 (38 μmol/L), and patient 5 the highest C10:0 concentration (88 μmol/L).

Hexanoylglycine (Reference < 4.53 mmol/mol creatinine) and suberylglycine (Reference < 9.87 mmol/mol creatinine) were increased in all the patients, while the dicarboxylic acid profile was normal in some of them. The heterozygous patients had also normal values of hexanoylglycine and suberylglycine

The diagnosis of MCADD was confirmed by molecular studies. Seven patients (12 alleles, as 2 are siblings) were homozygous for the prevalent c.985A > G (p.K329E) mutation. Three patients were compound heterozygous for the prevalent c.985A > G mutation and for another mutation, and one patient does not carry the c.985A > G mutation in any of the alleles (Table 1). Two of the mutations c.245G > C (p.W82S) and c.542A > G (p.D181G) are novel. The c.1247T > C (p.I416T) has very recently been reported in a US newborn identified by MS/MS-based routine screening (Smith et al. 2010), and we have furthermore observed this mutation in another US newborn identified by MS/MS-based routine screening (personal communication). Among our MCADD individuals (20 alleles), the prevalence of the common allele is 75%.

After diagnosis, all the patients followed the dietary recommendations described above. Cases 3 and 4 were supplemented with L-carnitine due to the low levels of free carnitine (below 12 μM). The evolutionary follow-up showed a normal physical and neurological outcome in all the patients, and until present ten have remained without clinical symptoms. Patient 4 died at 2 years of age in another hospital, due to a serious respiratory infectious, and we cannot identify whether the underlying disease influenced the outcome.

This patient was homozygous for the common mutation, and at diagnosis C8 and C8/C10 were strongly increased (Table 1). The remaining patients have never suffered any episode of decompensation.

All the patients have normal PDI/IQ, and those with the highest PDI/IQ are compound heterozygous for the common and another mutation (patients 8 and 9).

Three newborns are of Gypsy ethnicity and all are homozygous for the prevalent mutation, in agreement with the high prevalence of this mutation in this ethnic group.

Discussion

Although some children with MCADD may remain asymptomatic, even if they are not diagnosed, many papers highlight the importance of an early diagnosis and treatment to reduce the number of decompensation episodes and deaths (Nennstiel-Ratzel et al. 2005; Wilson et al. 1999). Early diagnosis of MCADD is now possible through the implementation of NBS by MS/MS. It allows the establishment of an early treatment and follow-up to prevent the acute metabolic derangement often associated with this disease. In addition, NBS by MS/MS also enables estimates of the prevalence of MCADD in the general population of a defined population to be made.

It has been reported that the incidence of MCADD is similar to that of phenylketonuria (1:10,000 live newborns) or even higher (Grosse et al. 2006). In our population, the frequency is 1/18,847 live births. Since our center receives samples from all the infants born in our area, the NBS results of MCADD reflect the characteristics of the Galician population.

As previously reported by others, we have also found that the combined elevation of C8 and C8/C10 ratio is the most useful tool for most of the MCADD diagnosis (Blois et al. 2005), but the ratio C8/C10 was normal in patient 8. According to Smith, this patient could be considered as having an intermediate phenotype and acylcarnitine results are in agreement with previous descriptions (Smith et al. 2010; Maier et al. 2005). The average value of C8 in our homozygous patients for the common mutation was 3.9 μmol/L, while the value decreased to 1.28 μmol/L in compound heterozygous. These results are similar to those reported by other groups (Al-Hassnan et al. 2010), but lower than those reported by Blois (Blois et al. 2005) with a median value of 13.8 μmol/L for homozygous and of 2.6 μmol/L for compound heterozygous.

Plasma FFA and urine acylglycine profiles are helpful for the diagnosis, but they are not essential for it (Onkenhout et al. 1995; Bonafé et al. 2000; Kobayashi et al. 2007). However, the clearly elevated concentration of cis-4-decenoic acid (C10: 1n-9) (Table 1) in patient 8 with very mild elevation of acylcarnitines and normal C8/C10 ratio, contributed decisively to the diagnosis.

Similar to other studies, the prevalent mutation in our population is c.985A>G (p.K329E), which was found in 75% of the alleles with MCADD, but was not as high as previously anticipated (Yokota et al. 1991; Giroux et al. 2007; Waddell et al. 2006; Blois et al. 2005), although it is rather high if we consider other neonatal screening studies (Andresen et al. 2001; Maier et al. 2005; Nichols et al. 2008; Smith et al. 2010) as in the screened individuals it is usual to find milder mutations, while the common mutation c.985A>G is most frequently found in the clinically presenting patients.

It seems that there is no clear relationship between genotype and phenotype (Andresen et al. 1997; Lehotay et al. 2004; Waddell et al. 2006; Hsu et al. 2008), but patients homozygous for the common mutation have higher levels of C8 (Andresen et al. 2001; Waddell et al. 2006; Smith et al. 2010), and it is associated with a greater predisposition to suffer decompensation in situations of metabolic stress (Arnold et al. 2010). Thus, compound heterozygous for the prevalent c.985A>G mutation and a milder mutation in the other allele may have a risk reduction due to their greater residual enzyme activity (Lehotay et al. 2004). Among our population patient 4, homozygous for the c.985A>G, mutation died at the age of 2 years in the context of a severe intercurrent illness. The remaining patients, including his brother, who is also homozygous for c.985A>G, have not suffered any episode of metabolic decompensation. Therefore, the control of environmental factors seems to be essential, as they could play an important role in the clinical development and in the expressiveness of the MCADD (Gregersen et al. 2008).

In addition to the common mutation, five other mutations have been detected in our group of patients. The c.799G>A (p.G267R) mutation has been previously reported in symptomatic patients (Yokota et al. 1991; Andresen et al. 1997; Zschocke et al. 2001). Glycine 267 is highly conserved in humans and other organisms and are found at the equivalent positions in human short-chain and branched-chain acyl-CoA dehydrogenases (Zschocke et al. 2001). Expression studies in *Escherichia coli* revealed a decrease of the enzyme activity that could be increased by co-overexpression of the GroESL chaperonins (Andresen et al. 1997).

Mutation c.199T>C (p.Y67H) found in patient 8, has also been previously reported, and is the second most prevalent mutation of MCADD (Andresen et al. 2001). This mutation has never been identified in patients with a clinical phenotype, but in several newborns identified by neonatal screening in the United States, Australia, and Germany (Maier et al. 2005; Smith et al. 2010; Zschocke et al. 2001). These newborns had a relatively mild acylcarnitine profile, like our patient 8, that was compound heterozygous for this and for the common mutation and only presented a slight elevation of metabolites. Expression studies of the p.Y67H mutation in *E. coli* revealed that the biogenesis and/or stability are compromised (Andresen et al. 2001), but the catalytic activity of the enzyme was minimally affected (O'Reilly et al. 2004). However, there is several reason to believe that this mutation is not clinically neutral, particularly at high body temperatures (O'Reilly et al. 2004).

In addition, two novel mutations c.245G>C (p.W82S) and c.542A>G (p.D181G) were identified. Expression studies of these mutations have not been done, but several lines of evidence suggest that they may be pathogenic: (1) the nucleotide changes were not present in more than 100 control chromosomes analyzed, (2) tryptophan in position 82 is a residue highly conserved among different species and among other acyl-CoA dehydrogenases (http://coot.embl.de/PolyPhen/), (3) in addition, p.W82S is homologous to the p.G101R mutation in GCDH protein, which may indicate that changes at this position of acyl-CoA dehydrogenase proteins are not tolerated. Concerning p.D181G mutation, its pathogenecity is uncertain and further studies are necessary to elucidate whether it could be disease causing.

From this study, and in agreement with other authors (Pollitt and Leonard 1998; Blois et al. 2005), we can conclude that NBS for MCADD must be strongly recommended due to its relatively high incidence coupled with an easy and specific detection and simple treatment.

Sypnosis

This study presents the first published data about neonatal screening for MCADD and evolution of the diagnosed patients in Spain (region of Galicia) describing the prevalence of mutations, two novel mutations and the correlation phenotype–genotype.

Ethical Statement

- All the authors have contributed equally to the planning and execution of this work.
- M.L. Couce serves as guarantor for the article, accepts full responsibility for the work and the conduct of the study, had full access to the data, and controlled the decision to publish.
- Competing interest statement: All authors declare that they have no competing interests.
- The authors confirm independence from the sponsors; the content of this article has not been influenced by the sponsors.
- The patients' parents were fully informed about this study and gave informed voluntary consent to participation.

The authors

References

Al-Hassnan ZN, Imtiaz F, Al-Amoudi M (2010) Medium-chain acyl-CoA dehydrogenase deficiency in Saudi Arabia: incidence, genotype, and preventive implications. J Inherit Metab Dis. doi:10.1007/s10545-010-9143-1

Andresen BS, Bross P, Jensen TG et al (1993) A rare disease-associated mutation in the gene for medium chain Acyl-CoA dehydrogenase (MCAD) changes a conserved arginine residue previously shown to be functionally essential in short-chain Acyl-CoA dehydrogenase (SCAD). Am J Hum Genet 53:730–739

Andresen BS, Bross P, Udvari S et al (1997) The molecular basis of medium-chain acyl-CoA dehydrogenase (MCAD) deficiency in compound heterozygous patients: is there correlation between genotype and phenotype? Hum Mol Genet 6:695–707

Andresen BS, Dobrowolski SF, O'Reilly L et al (2001) Medium-chain acyl-CoA dehydrogenase (MCAD) mutations identified by MS/MS-based prospective screening of newborns differ from those observed in patients with clinical symptoms: identification and characterization of a new, prevalent mutation that results in mild MCAD deficiency. Am J Hum Genet 68:1408–1418

Arnold GL, Saavedra-Matiz CA, Galvin-Parton PA et al (2010) Lack of genotype-phenotype correlations and outcome in MCAD deficiency diagnosed by newborn screening in New York State. Mol Genet Metab 99:263–268

Blois B, Riddell C, Dooley K, Dyack S (2005) Newborns with C8 acylcarnitine level over the 90th centile have an increased frequency of the common MCAD 985A>G mutation. J Inherit Metab Dis 28:551–556

Bonafé L, Troxler H, Kuster T et al (2000) Evaluation of urinary acylglicines by electrospray tandem mass spectrometry in mitochondrial energy metabolism defects and organic acidurias. Mol Genet Metab 69:302–311

Chace DH, DiPerna JC, Mitchell BL, Sgroi B, Hofman LF, Naylor EW (2001) Electrospray tandem mass spectrometry for analysis of acylcarnitines in dried postmortem blood specimens collected at autopsy from infants with unexplained cause of death. Clin Chem 47: 1166–1182

Giroux S, Dubé-Linteau A, Cardinal G et al (2007) Assessment of the prevalence of the 985A>G MCAD mutation in the french-canadian population using allele-specific PCR. Clin Genet 71:569–575

Gregersen N, Blakemore A, Winter V et al (1991) Specific diagnosis of medium-chain acyl-CoA dehydrogenase (MCAD) deficiency in dried blood spots by a polymerase chain reaction (PCR) assay detecting a point-mutation (G985) in the MCAD gene. Clin Chim Acta 203:23–34

Gregersen N, Andresen BS, Pedersen CB, Olsen RK, Corydon TJ, Bross P (2008) Mitochondrial fatty acid oxidation defects-remaining challenges. J Inherit Metab Dis 31:643–657

Grosse SD, Khoury MJ, Greene CL, Crider KS, Pollitt RJ (2006) The epidemiology of medium chain acyl-CoA dehydrogenase deficiency: an update. Genet Med 8:205–212

Hsu HW, Zytkovic TH, Comeau AM et al (2008) Spectrum of medium-chain acyl-CoA dehydrogenase deficiency detected by newborn screening. Pediatrics 121:108–114

Kobayashi H, Hasegawa Y, Endo M, Purevsuren J, Yamaguchi S (2007) ESI-MS/MS study of acylcarnitine profiles in urine from patients with organic acidemias and fatty acid oxidation disorders. J Chromatogr B Analyt Technol Biomed Life Sci 855:80–87

Lehotay DC, LePage J, Thompson JR, Rockman-Greenberg C (2004) Blood acylcarnitine levels in normal newborns and heterocygotes for medium-chan acyl-CoA dehydrogenase deficiency: a relationship between genotype and biochemical phenotype? J Inherit Metab Dis 27:81–88

Maier EM, Liebl B, Röschinger W et al (2005) Population spectrum of ACADM genotypes correlated to biochemical phenotypes in newborn screening for medium-chain acyl-CoA dehydrogenase deficiency. Hum Mutat 25:443–452

Martínez G, Jiménez-Sánchez G, Divry P et al (1997) Plasma free fatty acids in mitochondrial fatty acid oxidation defects. Clin Chim Acta 267:143–154

Millington DS, Kodo N, Norwood DL, Roe CR (1990) Tandem mass spectrometry: a new method for acylcarnitine profiling with potential for neonatal screening for inborn errors of metabolism. J Inherit Metab Dis 13:321–324

Nennstiel-Ratzel U, Arenz S, Maier EM et al (2005) Reduced incidence of severe metabolic crisis or death in children with medium chain acyl-CoA dehydrogenase deficiency homozygous for c.985A>G identified by neonatal screening. Mol Genet Metab 85:157–159

Nichols MJ, Saavedra-Matiz CA, Pass KA, Caggana M (2008) Novel mutations causing medium chain acyl-CoA dehydrogenase deficiency: under-representation of the common c.985A>G mutation in the New York state population. Am J Med Genet A 146:610–619

Onkenhout W, Venizelos V, van der Poet PFH, van den Heuvel MP, Poorthuis BJ (1995) Identification and quantification of intermediates of unsaturated fatty acid metabolism in plasma of patients with fatty acid oxidation disorders. Clin Chem 41:1467–1474

O'Reilly L, Bross P, Corydon TJ et al (2004) The Y42H mutation in medium-chain acyl-CoA dehydrogenase, which is prevalent in babies identified by MS/MS-based newborn screening, is temperature sensitive. Eur J Biochem 271:4053–4063

Pollitt RJ, Leonard JV (1998) Prospective surveillance study of medium chain acyl-CoA dehydrogenase deficiency in the UK. Arch Dis Child 79:116–119

Smith EH, Thomas C, McHugh D et al (2010) Allelic diversity in MCAD deficiency: the biochemical classification of 54 variants identified during 5 years of ACADM sequencing. Mol Genet Metab 100(3):241–250, Epub 2010 Apr 8.PMID: 20434380 [PubMed – in process]

Tanaka K, West-Dull A, Hine DG, Lynn TB, Lowe T (1980) Gas-chromatographic method of analysis for urinary organic acids. II. Description of the procedure, and its application to diagnosis of patients with organic acidurias. Clin Chem 26:1847–1853

Tran K, Banerjee S, Li H, Noorani HZ, Mensinkai S, Dooley K (2007) Clinical efficacy and cost-effectiveness of newborn screening for medium chain acyl-CoA dehydrogenase deficiency using tandem mass spectrometry. Clin Biochem 40:235–241

Waddell L, Wiley V, Carpenter K et al (2006) Medium-chain acyl-CoA dehydrogenase deficiency: genotype–biochemical phenotype correlations. Mol Genet Metab 87:32–39

Wilcken B, Haas M, Joy P et al (2007) Outcome of neonatal screening for medium-chain acyl-CoA dehydrogenase deficiency in Australia: a cohort study. Lancet 369:37–42

Wilson CJ, Champion MP, Collins JE, Clayton PT, Leonard JV (1999) Outcome of medium chain acyl-CoA dehydrogenase deficiency after diagnosis. Arch Dis Child 80:459–462

Yokota I, Coates PM, Hale DE, Rinaldo P, Tanaka K (1991) Molecular survey of a prevalent mutation, 985A-to-G transition, and identification of five infrequent mutations in the medium- chain acyl-CoA dehydrogenase (MCAD) gene in 55 patients with MCAD deficiency. Am J Hum Genet 49:1280–1291

Zschocke J, Schulze A, Lindner M et al (2001) Molecular and functional characterisation of mild MCAD deficiency. Hum Genet 108 (5):404–408

MIX
Papier aus verantwortungsvollen Quellen
Paper from responsible sources
FSC® C105338

If you have any concerns about our products,
you can contact us on
ProductSafety@springernature.com

In case Publisher is established outside the EU,
the EU authorized representative is:
Springer Nature Customer Service Center GmbH
Europaplatz 3, 69115 Heidelberg, Germany

Printed by Libri Plureos GmbH
in Hamburg, Germany